THINKING
ABOUT
INNOVATION

THINKING
ABOUT
INNOVATION

How Coffee, Libraries,
Western Movies,
Modern Art, and AI
Changed the World of Business

DR. ROGER D. SMITH

M

Modelbenders Press

Thinking About Innovation: How Coffee, Libraries, Western Movies, Modern Art, and AI Changed the World of Business

Modelbenders Press books may be purchased for business and promotional use or for special sales. For information please contact the publisher.

PRINTED IN THE UNITED STATES OF AMERICA

Visit our web site at www.modelbenders.com

Designed by Adina Cucicov at Flamingo Designs
Cover image: © Kanisis | Dreamstime.com

The Library of Congress has cataloged the paperback edition as follows:

Smith, Roger D.
    Thinking About Innovation: How Coffee, Libraries, Western Movies,
    Modern Art, and AI Changed the World of Business
    Roger D. Smith.—1st ed.
    1. Innovation & Creativity 2. Business & Economics: Leadership
    3. Business & Economics: Research and Development
    4. Business & Economics: Industrial Management
    I. Roger D. Smith II. Title

ISBN 978-1-9385900-8-5

ALSO BY ROGER SMITH

Chief Technology Officer:
Defining the Responsibilities of the Senior Technical Executive

TABLE OF CONTENTS

FOREWORD

I n 2008 the Editor of *Research Technology Management* asked if I would be interested in writing a regular column on a topic of interest to industrial research managers. As a practicing CTO I was regularly challenged to solve specific problems within our projects, identify ways to optimize our business practices, and search for competitive advantages that would differentiate us from the competition. This made me an avid reader of books on leadership, change management, creativity, and innovation. For the column I needed to combine lessons from these books, real problems from within the company, and experiences from creating new solutions. All of this had to be done without revealing proprietary information about our internal business strategies, and it had to be done on a publishing schedule.

That column has now run for over ten years and has become the most popular item in the journal. During that time I changed jobs several times, but remained in a technology leadership position like Technical Director, VP of Technology, Chief Scientist, or Chief Technology Officer, the latter being the most common. Through my 12 years in a CTO-like role I have constantly struggled to stimulate innovation at work and to describe or prescribe new practices in the RTM column.

Those columns and several additional papers are collected here for the benefit of other leaders who are faced with similar challenges. From experience I know that no two companies or leaders face exactly the same problems. But the ideas in these chapters will inspire and equip you to handle your unique situations more effectively.

I remain ... *A CTO Thinking About Innovation*.

PREFACE

THE SECT OF INNOVATION ...

It is customary for books on innovation to offer a definition of the term somewhere in the opening chapters. But, by now, every reader has seen the few leading definitions and will likely skip over their repeated appearance. Readers typically come to a book on innovation with their own internal definition of this broad and somewhat vague term. In fact, many of them consider it a form of innovation to fiercely maintain a personally unique and even contradictory definition of innovation. Therefore, I will refrain from attempting to explain what innovation is, assuming that the reader already has a working definition that they are quite happy with.

Through the 1990's and 2000's the cult of innovation grew very robustly. Today, there are at least 3,000 business books on the subject of innovation with nearly 300 being added each year. It seems that the members of the innovation cult have a voracious appetite for innovative ideas about innovation. The flock and their priestly gurus aggressively proselytize the concept with the hope of winning millions to the cause. Those being inducted into the faith read the scriptures on innovation and become concerned for the soul of their business. Should they refuse to convert to innovation-ism, they fear that the entire organization will be disrupted and left on the trash heap of history. So most accept the faith, and work to incorporate its doctrines into their daily business practices.

By the second decade of the 21st century, every leader, executive, manager, academic, student, and aspirant has heard of innovation and has added it as an adjective to their professional profile, resume, and mission statement. Inwardly, we all fear that we might not be changing fast enough and some super-innovator will emerge from behind, zoom past us, and snatch away our promotion, raise, or business.

Definitions from the Inspired Book of Innovation:

Innovation 1:1—A new idea, method, or device (Merriam-Webster.com)

Innovation 1:2—The process by which an idea or invention is translated into a good or service for which people will pay, or something that results from this process. To be called an innovation, an idea must be replicable at an economical cost and must satisfy a specific need. (BusinessDictionary.com)

Innovation 1:3—The creation of better or more effective products, processes, services, technologies, or ideas that are accepted by markets, governments, and society. Innovation differs from invention in that innovation refers to the use of a new idea or method, whereas invention refers more directly to the creation of the idea or method itself. (Wikipedia.org)

I expect that most readers of this book are not hearing about innovation for the first time, but rather are seeking new and stimulating ideas on the subject which will help them maintain an open, inquisitive, and creative mind.

Business practitioners will probably come to this book having already read the excellent works of Clayton Christensen, Henry Chesbrough, Eric von Hipple, Andrew Hargadon, CK Prahalad, W. Chan Kim, Renée Mauborgne, Richard Leifer, Edward Roberts, and others.

Academic researchers will have read the papers of Burns & Stalker, Utterback, Rogers, Tushman, Katz, Leonard-Barton, Sharif, and dozens like them.

All of these works have inspired and directed the ideas and practices presented in this book. All of those authors have made a difference in the business practices and academic writings of hundreds of people, including my own.

THE PRACTICE OF INNOVATION ...

My career as an innovator began long before the term was widely used or revered. I emerged from college (the first time) with degrees in mathematics and statistics and found myself conducting research into new applications of national security systems. As a mathematician I was expected to create new algorithms to describe the physical and conceptual behavior of aircraft, weapons, and their targets on the ground. I was an innovator and did not know it yet.

In the late 1980s' and early 1990's all mathematicians and scientists who could type were swept into the ranks of computer programmers, both eagerly and against their wills. The demand for computerization of every aspect of the world turned most scientific jobs into full-time or part-time programming jobs. In this role, I found myself creating computer software to describe the behaviors of systems in the real world that had never before been captured in a computable form. I had been swept into yet another form of innovation.

My success in creating software and in writing business proposals to capture new clients led to multiple promotions through "Senior this" and "Director that", until I found myself with the title of Chief Technology Officer. A review of the literature revealed that there had been very little work to define what a CTO was and what one was expected to do. But, the dot.com era called for a C-suite position for the geniuses who created the products, so every company had to have one. While exercising my CTO responsibility to identify new technologies that could have a substantial positive or negative impact on our business, I also exercised the responsibility of every Ph.D. to conduct research and publish new knowledge. In my case, I chose to spend a few years researching and defining the role of the CTO. Perhaps this made me a double innovator for a short time.

Finally, I arrive in the executive ranks with the responsibility to lead parts of the organization into the undefined and unknowable future, showing both confidence and insight to inspire the crew to new achievements. We can either be innovative or imitative with our products and services. The former promises recognition, financial reward, and organizational success. The latter threatens stasis, poor financial margins, and a slow decline. So we struggle to convince everyone that we must innovate or die, when in truth many would often rather die than innovate.

INSIGHTS ON INNOVATION ...

After 20 years of practicing innovation in science, technology, management, and leadership I believe that I have learned a few principles of successful innovation. The wisdom that I have gained from dozens of books and papers on the subject has also guided my practice and perspective on this important topic. The essays in this book illustrate many of those beliefs and experiences. There are a few key characteristics of innovation that aspiring practitioners should be aware of as illustrated in the essays. To insure that those characteristics are not missed, I will express some of them succinctly here.

- Innovation is always wished for, but generally not welcome when it arrives.
- Innovation generally occurs at the edges where two disciplines intersect.

- Innovation is difficult to sustain in any organization, either large or small.
- Innovation is driven by personal passion, talent, and initiative.
- Innovation is birthed by the young and old alike, age is not a barrier.
- Innovation is threatening when first met face to face.
- The birth of an innovative idea can be very messy, ugly, and terrifying.
- Innovation is appreciated only after it passes through its adolescent phase.

PART I

INNOVATION CULTURE

CHAPTER 1

THE ELIXIR OF INNOVATION

In his famous diaries, Samuel Pepys, the 17th-century London author, described his daily repast: Each morning he would break his fast with a pint of beer or ale. At mid-day, he would lunch on meat, bread, and a pint of either beer or wine. On the way home in the evening, he would stop for ale or hard liquor. Dinner included at least another pint of beer.

P epys's menu was not unusual. Entire countries were awash in an alcoholic stupor. In the majority of the population, this continuous sotting of the brain engendered a lethargy of body and mind that suppressed productivity and dulled creativity. Even a hundred years later, the effects were evident; Benjamin Franklin describes the situation as he found it in the London printing house to which he was apprenticed in 1725:

> I drank only water; the other workmen, near fifty in number, were great guzzlers of beer. On occasion, I carried up and down stairs a large form of types in each hand, when others carried but one in both hands. They wondered to see, from this and several instances, that the Water-American, as they called me, was stronger than themselves, who drank strong beer!
>
> We had an alehouse boy who attended always in the house to supply the workmen. My companion at the press drank every day a pint before breakfast, a pint at breakfast with his bread and cheese, a pint between breakfast and dinner, a pint at dinner, a pint in the afternoon about six o'clock, and another when he had done his day's work.

Europe found its solution to this problem in 1652, with the introduction of a wondrous new elixir that came to London in the hands of an Armenian named Pasqua Rosee. Rosee opened a new kind of pub in St. Michael's Alley, serving a Turkish brewed beverage known as *kaweh*, a word that translates as "strength and vigor." Today, we are more familiar with the English adaptation of that word, coffee. From that single café, the new drink, previously little known in Europe, grew rapidly in popularity: there were more than 3,000 coffeehouses in England by 1675, just 23 years later. This was an expansion on the order of Starbucks' growth in the 1990s.

Pasqua Rosee himself benefited from the new business, most likely opening multiple houses in London and other parts of England. In 1672, he became an international franchiser when he opened the first coffeehouse in Paris. Rosee enjoyed a citywide monopoly on the business across Paris for 14 years, until his first competitor opened in 1686.

When he opened his Paris coffeehouse, Rosee may not have been thinking of anything larger than expanding his own personal wealth, but his business had huge effects—it arguably became a pillar of the nascent French Enlightenment. Rosee's house was one of the meeting places for French Enlightenment thinkers like Voltaire, Rousseau, and Diderot, who is credited with creating the first modern encyclopedia in that coffeehouse.

But certainly, caffeine cannot not singlehandedly transform a country into an innovation machine. If it were that simple, every country would just import stimulants to trigger the creation of new ideas and new businesses and buoy their economies. It was certainly important that France's great thinkers weren't wine-addled, but coffee, and the coffeehouses in which it was served, provided other important supports, nourishing individual creativity in less obvious ways by providing a community and a culture that sparked ideas and supported innovation

COMMUNITY AND CULTURE

Coffeehouses provided both the meeting places that brought great minds into contact—allowing their ideas to collide and grow with unprecedented productivity—and the fuel for their discussions. Suddenly, intellectuals across the continent, chemically stimulated by coffee, were engaging in vigorous political discussion, advancing philosophy, and creating new schools of art—and entire new industries: the insurance industry was born with the creation of Lloyd's of London in a coffeehouse in 1688, the London Stock Exchange formed in one in 1698, and Sotheby's and Christie's were each formed in coffeehouses, in 1744 and 1759, respectively. Similarly, Wall Street (the New York Stock Exchange), which was famously started under a buttonwood tree in 1792, found its first home inside the Tontine Coffee House just a few months later.

Lloyd's of London is a particularly interesting case. Edward Lloyd was originally a coffeehouse owner. Located near the Tower of London, where shipping businesses converged, Lloyd's Coffeehouse became a meeting place for ship owners, merchant exporters, and sailors. Presiding over this convergence of people with a common need gave Lloyd insight into the need for a more comprehensive way to insure ships and their cargo. Thus, Lloyd transformed himself from a coffeehouse proprietor to a broker, connecting those with a need for insurance and those with the financial means to provide it. Over time, he built Lloyd's of London into an insurance market within which wealthy individuals and businesses could offer insurance to the shipping industry. This market approach to pooling resources and spreading risk proved more stable than any individual insurance company, whose limited resources put it at constant risk of being wiped out by a few large losses.

From a certain perspective, the meeting place may have been more important in sparking Lloyd's innovation than the coffee itself. But the coffee provided both the reason for assembly and the fuel for long hours of discussion and problem solving. The wealthy virtuosi, who were expected to pursue learning for its own sake, might meet and exchange ideas at a salon or a scientific society meeting. But the merchants and craftsmen, shrewd innovators with a keen sense for a practical opportunity, had only the coffeehouses.

Other times and places also saw coffee-fueled bursts of innovation. In the 1960s, American folk music, and artists like Bob Dillon and Joni Mitchell, found a place to mature and build a following in coffeehouses. Those coffeehouses nurtured the social revolution that would come to define the period, and remake American society.

Similarly, the explosive wave of innovation that put Silicon Valley on the innovation map in the 1990s erupted almost symbiotically with the emergence of Starbucks and Pete's Coffee & Tea, and their new dedication to coffeehouse culture.

CONCLUSION

Often, innovation is not driven by the brilliant insights of a single person or small group. Rather, it emerges from the clash of many ideas that occur in a social space, like a coffeehouse. In cases like those cited here, that productive exchange is enhanced by the chemical stimulant of the coffee itself, the social stimulant of the coffeehouses, and the cultural stimulants engendered by people's encounters in the coffeehouses.

With these historical cases in mind, we might ask ourselves, where is the coffeehouse in my company? What am I doing to create and nourish an innovation community where I work? Are the chemical, social, and cultural stimulants needed for innovation accessible to everyone in my company?

Originally Published in *Research Technology Management*, Jan-Feb 2015

CHAPTER 2

LEONARDO: BRIDGING THE GAP

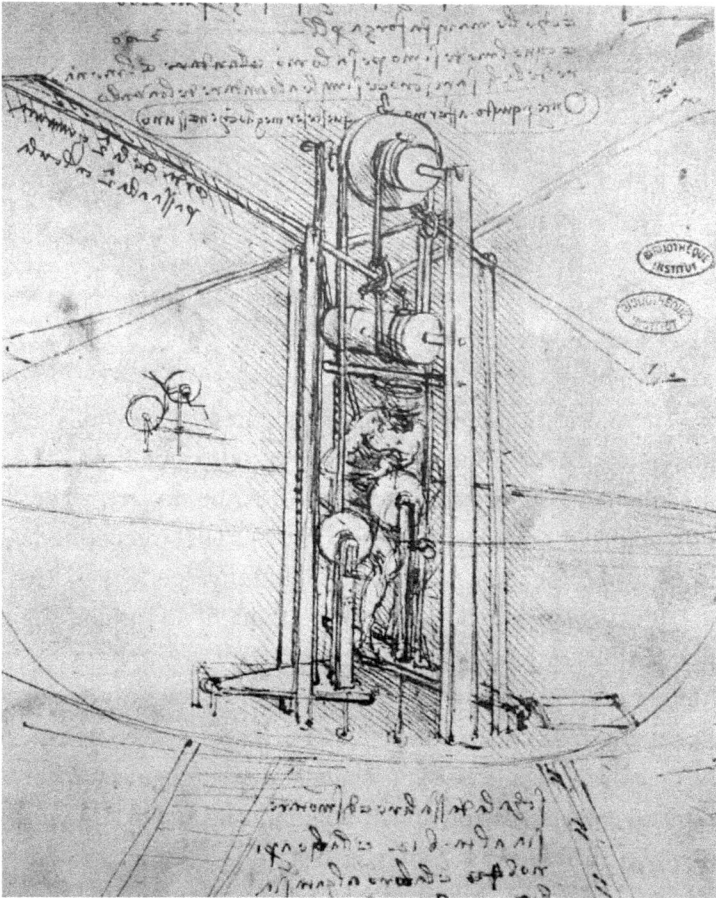

T he history of science is filled with the lives and legends of men and women who have made huge leaps forward in our knowledge and understanding. These people have made contributions that reverberate for centuries and their names have become synonymous with the science and the ideas that they created:

> Copernicus, 1514: "The sun does not move. The earth is not in the center of the circle of the sun, nor in the center of the universe."

> Galileo, 1609: "A large magnifying lens should be employed to study the surface of the moon and other heavenly bodies."

> Isaac Newton, 1679: "Every weight tends to fall towards the center by the shortest possible way."

> Charles Darwin, 1856: "Man does not vary from the animals except in what is accidental."

These are the names we have attached to each of these world-changing ideas, but none of the quotations came from the men who are credited with the concepts they describe. In fact, all are from the notebooks of Leonardo da Vinci (1452–1519), predating by decades or centuries the great men who would fully define the ideas.

How can any single person have thought so insightfully about so many different areas of science?

Leonardo of the city of Vinci in Italy was one of the most prolific polymaths recorded in human history. He was an accomplished painter, sculptor, architect, musician, mathematician, engineer, inventor, anatomist, geologist, cartographer, botanist, and writer. He worked as an artist for the Medici in Florence, Corvinus in Hungary, and Pope Leo X; a civil engineer for Ludovico in Milan and Ottoman Sultan Beyazid

II in Constantinople; a military architect in Venice; a cartographer for the family of Pope Alexander VI; an anatomist with Marcantonio della Torre in Pavia and Padua; and finally, as a military and civil engineer for Francis I of France, who captured Milan and took Leonardo back to France as a valuable trophy of war.

He was able to jump between all of these fields to make valuable contributions when they were still young sciences. He lived at a time when the social and professional boundaries between the disciplines were just beginning to be erected; it was not clear where one field ended and the other began, so there was nothing to prevent an agile mind from crossing back and forth between them.

Leonardo was certified as a master in the Guild of St. Luke— which credentialed both artists and doctors of medicine—at the age of 20, after a six-year apprenticeship, but he was not a man to define himself by the boundaries of a guild. For most people, early training in a specific guild would define their life's work, focusing their contributions in a single field. For Leonardo, art was just a foundation, a tool for growing into other areas. He bridged the gap from one profession to another when it suited his curiosity and his insights. In his own words,

> I roamed the countryside searching for answers to things I did not understand. Why shells existed on the tops of mountains along with the imprints of coral and plants and seaweed usually found in the sea. Why the thunder lasts a longer time than that which causes it, and why immediately on its creation the lightning becomes visible to the eye while thunder requires time to travel. How the various circles of water form around the spot which has been struck by a stone, and why a bird sustains itself in the air. These questions and other strange phenomena engage my thought throughout my life.

Amazingly, he seems to have been unique in this exploration. Historical records do not include drawings and inventions by other artists intent on changing the world. It seems that his peers stuck to their painting and left engineering and inventions to the craft guilds of carpentry, waterworks, and armorers.

By contrast, Leonardo seems to have deployed his formal training as an artist to support his scientific imagination. For instance, his concept for a hydraulic screw for lifting water from lower to higher heights required an understanding of physics and engineering, but also the ability to draw the device in detail, to illustrate it for the patrons who would fund the project and the craftsmen who would build it. Verrocchio, the master to whom the young Leonardo was apprenticed, taught his students to observe objects and nature in detail, to understand how bodies and natural objects move and interact.

The powers of observation these exercises developed in the young Leonardo became the tools of a scientist and engineer. They allowed him to understand physics and natural science in a practical way, so he could design innovative new machines. This ability to see and understand the world, rooted in his artistic training, led to entire treatises on anatomy, the flight of birds, and the construction of machines.

Leonardo understood, as few before or since have, that his mind could only remain sharp if he used it constantly and aggressively. "Iron rusts from disuse," he wrote; "stagnant water loses its purity and in cold weather becomes frozen; even so does inaction sap the vigor of the mind." He recognized the full, rich set of abilities of his mind and refused to let any part of it rust from disuse. No one guild or profession could fully sustain his growth or contain his intellect, so he ranged beyond his official, credentialed field and made lasting contributions in multiple fields. He reached the end of his life satisfied that he had done everything he could to leave a lasting contribution for the world.

The guild system Leonardo so deftly defied eventually evolved into the universities we have today. The boundaries between disciplines that were vague 500 years ago have become high-walled cities that can seem impenetrable. Scientists find themselves almost completely confined to a single discipline by their focused education, isolated professional communities, and organizational units—whether company divisions or university departments—that recognize only one kind of contribution. But in a world where most single-discipline problems have been solved, the big payoffs are in solutions to multidisciplinary problems that call for individuals and teams who can integrate the skills and perspectives of many fields. Real value comes from people who can build within themselves the skills and capabilities required to approach these new problems.

What are needed today are modern Leonardos, individuals who can extend themselves beyond their formal training and integrate skills to match the diversity of the difficult problems to be solved. Universities, companies, professional organizations, and individuals themselves need to promote and pursue the development and integration of all of the diverse talents that are latent in each person. In a modern world with more knowledge and more people than has ever existed, there should be hundreds or thousands of Leonardos, making contributions across multiple fields and leaving the world a better place for their having escaped the artificial boundaries of discipline and division.

Originally Published in *Research Technology Management*, Jan-Feb 2014

CHAPTER 3

WILD WEST INNOVATION

"You see, in this world there's two kinds of people, my friend: Those with loaded guns and those who dig. You dig."

- CLINT EASTWOOD, *The Good, the Bad and the Ugly*

The Western movie has been a part of American culture for over a century, almost since the invention of the movie projector. Such an established form of entertainment would seem to be outside the need for business innovation. But it is precisely because the genre is so well trodden that innovation has always been an essential ingredient, producing new plots and new leading cowboys that keep the genre fresh.

The plot and hero of the western movie have been in constant evolution, with a number of very clear disruptors that shifted the entire genre punctuating decades of incremental change. Just as the hero of many a Western film rode into town and overturned the established power brokers, new actors regularly rode into Hollywood and changed the established hierarchy of the Western movie business. These changes on the screen and on the back lot are characteristic of the disruption that "innovation cowboys" like Steve Jobs and Mark Zuckerberg have had on established markets. And today's innovators can take some important lessons from these silver screen disruptors.

WESTERN INNOVATION

World famous innovator Thomas Edison is widely known as the creator of many of the technologies that led to the motion picture projector, but few know that he also owned the Edison Manufacturing Company, which produced and distributed movies to create commercial demand for the projector. His company created 1,200 films, beginning with short clips and moving on to full-length movies. Some of its earliest hit titles were *The Life of an American Cowboy* (1902) and *The Great Train Robbery* (1903), two of the first Western movies ever made (Matthews 1980). In the years that followed, producers and actors created various images of the cowboy hero to appeal to the general public, making cultural icons of men like "Bronco Billy" Anderson, the star of the *The Great Train Robbery*, who appeared in 466 short films between 1902 and 1922, and Bill Hart, who brought a range-rough image that included shooting up the bad guys and hard drinking to 74 films from 1907 to 1925. Both of these stars were created from the rough clay of their own personalities and their own ideas about what a cowboy was and did. With more than 500 movies carrying their images to the American public, they created the Hollywood version of this American hero.

But this image has never been set in stone, secure against change. Rather, the cowboy image has been open to regular re-creation, innovation, and disruption. New actors brought new styles to the screen, unseating reigning stars and redefining the Western. Hart's rough-clothed cowboy just off the range was unseated by the newcomer Tom Mix, a clean-shaven, fancy-dressed cowboy who preferred to rope the bad guys rather than shoot them dead. Mix created the "dandy cowboy," a well-groomed, clean-living hero who looked good on the screen but was a far reach from Hart's range cowboy.

Mix's career was in its latter years when a notable extra appeared on the set of *The Great K&A Train Robbery* (1926)—a football player from the University of Southern California who looked and acted the part of a strong onscreen cowboy. John Ford, the director of the film, invited Marion Morrison, or "Duke," as he was known to friends, to appear in several movies and eventually cast him as the lead in *The Big Trail* (1930) under his new Hollywood name, John Wayne. Wayne did not wear fancy rhinestone suits, polished silver, or spotless chaps. His image was reminiscent of Bill Hart's rough range cowboy, but without the violent vices. This update of the rough-and-ready cowboy became the face of the Western for the next 30 years.

But, there was room for more than one type of cowboy on the silver screen. The 1940s saw the introduction of an innovative new idea in Western entertainment, the singing cowboy. This soothing style was extremely popular in the years following the Great Depression; the leaders in the genre, Gene Autry and Roy Rogers, each starred in nearly 100 movies and television shows. Autry and Rogers also expanded the market for their characters, moving into radio programs, music records, and toys, pushing this innovative and profitable idea further than any previous stars had been able to.

Even the great John Wayne was not immune to change. In 1964, a bit actor in the *Rawhide* television series was eager to star in his own movie. But Hollywood did not see him as a cowboy who could carry on in the image of the Duke and declined to sign him up as a leading man. Italian and Spanish audiences liked their cowboys rougher and quieter, however, and directors there thought Clint Eastwood could deliver just the image that their audiences wanted. Eastwood's laconic presentation, along with a plot that returned the bloodshed and violence of earlier Westerns, made *A Fistful of Dollars* (1964) an instant success in Europe. The "Spaghetti Western," born in Italy, was eventually imported to America and eagerly embraced by audiences. Wayne's clean-cut hero was replaced by the rough outlaw.

LESSONS FROM COWBOY INNOVATORS

Together, the actors in front of the camera and the directors behind the scenes are involved in new product development to meet the changing needs and tastes of the market audience. These changes in the Western movie mirror those that occur in other businesses and offer effective lessons for initiating and surviving disruption.

Lesson 1: Innovators matter. A Western is significantly dependent upon a leading actor who can convincingly and consistently present an appealing character to the audience. Bill Hart, Tom Mix, Gene Autry, John Wayne, and Clint Eastwood all had the ability to portray unique characters that attracted audiences. Without these innovative leaders, the Western movie would have faded into the sunset decades ago.

Lesson 2: New-product success takes time. John Wayne appeared as an extra in Tom Mix movies and another 70 low-budget films without being noticed. Not until nearly his seventy-fifth movie, Stagecoach (1939), did he emerge as the icon that would carry the genre for 30 years.

Lesson 3: Niche products can succeed. Even while John Wayne was dominating the Western movie, there was a strong and loyal interest in singing cowboys like Gene Autry and Roy Rogers. Every market has fragments and each of these creates an opportunity for a unique product.

Lesson 4: Markets change. The Duke occupied the center of the Western stage for 30 years. But even before he was finished, audiences were turning to the new style of Clint Eastwood. Eventually, the audience will look for something new, no matter how dominant the current hero. Producers and directors need to remain in touch with the audience and be prepared to shift gears when the market changes.

Lesson 5: Foreign markets reveal new trends. Hollywood was firmly behind John Wayne, but audiences in Italy and Spain showed a strong interest in the outlaw image of Clint Eastwood. His success in those countries spurred Hollywood to test the image in America, creating a huge success and making room for the bad guy as a Western hero.

CONCLUSION

The Western movie has been one of the foundations of American entertainment and culture for over 100 years. Its heroes and ideals are embedded in the psyche of America and other countries. The movie cowboy portrays an image of a lone hero who breaks the rules but represents the morals and aspirations of every man in the audience. The construction of the modern Western movie and cowboy hero has been a story of disruptive innovation and change driven by individual innovators and risk takers. Even within the seemingly limited boundaries of the Western story, there is constant change in the market with a need to innovate to remain at the top.

References

Matthews, L. 1984. *History of the Western Movie. London: Hamlyn Publishing Group.*

Originally Published in *Research Technology Management*, Mar-Apr 2013

CHAPTER 4

THE RAINBOW ZEBRA

There is an old African legend about the origin of the zebra. At the beginning of time, the story goes, god was creating all of the animals on the grasslands. With great care, he crafted the beautiful gazelles, the sly hyenas, the elegant giraffes, and the proud lions. But he wished for one animal that was as colorful as the blooming flowers, the coral fish, and the shimmering birds. With these in mind, he created a zebra with stripes that were every color of the rainbow.

But the other animals were jealous of the zebra's bright colors. They thought the zebra was too haughty to live on the grasslands with them. The proud lions decided to hunt, kill, and eat all of the rainbow zebras. The zebras were almost completely gone when one of them called to god and asked him to save them from the lions. God was sad that the spectacular colors he had given this beautiful animal would be gone from the world, but to save the zebras' lives, he turned their stripes to plain black and white. Then he made them to run faster so the lions could not catch them easily. This satisfied the animals, and there was peace on the grasslands.

ZEBRAS AT WORK

The legend of the zebra speaks to the behaviors of people in their personal and work lives. Inside all of us, there is a beautiful and complicated pattern of colors. We are intellectual, emotional, physical, and spiritual—filled with unique aspirations, interests, and passions. All of these seek to express themselves in our lives without respect to the arbitrary, external boundary between work and personal spaces. At work, we want to be the creative and clever people that we are at home and with friends. But for over a century, the work tradition has been that each person expresses only two basic colors—the black stripe of our education and the white stripe of our job assignment. At work, the other colors inside us had to be muted, left dormant, only to be explored and expressed on our own time. Rainbow zebras couldn't survive.

This uniformity worked very well when most businesses were factories turning out standardized products day after day, year after year. Most employees only needed their education and assignment stripes to work efficiently in the factory. This fostered a productive organizational machine for repetitive work.

But today's leading companies must create new products and improve on existing ones at a much more rapid pace than in the past. A child of the 1960s might remember the introduction of one new brand of shampoo, one new potato chip, and one new soda during their entire childhood. But a child of the 2000s sees ten new shampoos every year, dozens of new snack chips, and hundreds of soft drinks. In this compressed, accelerated environment, the two traditional stripes of education and job assignment are no longer enough. Black and white zebras cannot create this much variety so quickly. Two-color zebras do not have access to all of the inner richness required to create a new product every week. They need a more complete and more varied set of stripes. It's time to bring back the rainbow zebra.

ZEBRA FREEDOM

In the 21st century, the opportunities we have and the problems we face require thousands of unique people who can apply unique combinations of talents to many different problems. They require people who are free to think and act differently to solve problems. Two-color zebras have access only to the solutions suggested by their education, training, and job role. Rainbow zebras can imagine, and try, thousands of different solutions. Many of them will fail, but that is the only way to find those that will succeed. Creating a rainbow of new products and innovative ideas requires the expression of the full rainbow of colors that reside in every person.

Companies like Google are famous for encouraging rainbow colors among their employees. Even a company of geniuses finds it necessary to encourage its engineers to spend 20 percent of their work time on ideas of their own choosing. The company also provides free food at on-campus restaurants; maintains massage facilities, gyms, and concierges; and provides support for continuing education. All of these are investments to manage, promote, and encourage their rainbow zebras. Few companies can afford all of the perks that Google uses to bring out the best in its people, but most organizations spend exactly zero on this kind of development. Can these organizations accomplish even a fraction of Google's results when they are investing nothing at all in promoting rainbow zebras?

Even for companies without the budget for free cafeterias, there are a number of affordable and approachable practices that have proven to stimulate innovative new ideas for organizations. These often focus on exposing people to new fields and new experiences that will allow them both to apply what they already know in new ways and to absorb new perspectives and solutions from outside their specialty. Some simple practices that promote rainbow zebra behavior include:

- Ask people to incorporate 15–30 minutes of reading on technical or management topics into their daily schedules.

- Encourage employees to socialize across departments, for instance by creating multidisciplinary lunch groups, so they interact with others who have different skills and experiences.

- Give occasional assignments outside of employees' specialty areas to stimulate them to grow in new areas and more importantly to motivate them to wrestle with unusual problems and develop new ways of thinking.

- Allow employees time to explore the organization so they see and hear what happens outside their usual routines and get to walk in other zebras' shoes.

- Pay attention to the role physical fitness plays in creative and mental work; the mind and imagination are chemical machines that are directly tied to the physical functions of the entire body.

Each of these practices has demonstrated benefits, and they require very few resources to implement. Some companies are thinking bigger, reaching out to other institutions for help in nurturing their rainbow zebras. BTG, a mid-sized defense contractor, recognized the importance of its zebras developing new skills constantly. Like other companies, it maintained a college tuition reimbursement program and encouraged its people to seek advanced degrees. Ed Bersoff, the CEO, understood that the level of employee expertise was directly linked to the company's win rate for important contracts. Acting on this belief, he worked closely with a large university to help it craft an MBA program that would meet the needs of his own and other similarly positioned companies. This emphasis on learning contributed to the company's growth, which in turn made it an attractive acquisition target; it is now part of a much larger corporation.

What does your company do to encourage rainbow-zebra diversity and richness of thinking, action, and interaction among employees? What more can it do? Every company needs to explore what will be effective in its business and determine what is affordable in both time and money. Doing nothing will not create exceptional people, exceptional products, exceptional profits, or exceptional competitive advantage. Does your company have a plan for developing the creativity, innovative abilities, and return on investment from its employees?

Every color of the rainbow that is suppressed is a lost resource. Advanced countries are evolving toward business practices that support rainbow zebras. We understand that the black and white patterns of the past century are not sufficient to promote and fuel the world into a colorful new future. George Bernard Shaw said, "The reasonable man adapts himself to the world; the unreasonable one persists in trying to adapt the world to himself. Therefore all progress depends on the unreasonable man." We need more unreasonable rainbow zebras. What are you doing to nurture the ones you have?

Originally Published in Research Technology Management, Sept-Oct 2013

CHAPTER 5

INNOVATE LIKE IT'S 1985

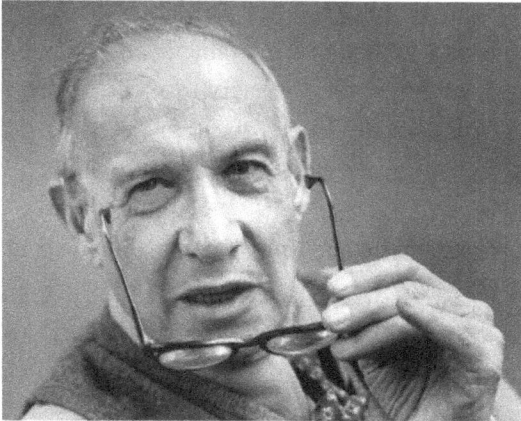

Courtesy of The Drucker Institute at Claremont Graduate School

nnovation has become a cornerstone of all growing businesses. We innovate on product, packaging, delivery, service, color, scent, "mouth feel," and every other feature a potential customer can experience. We innovate just as vigorously on the methods of innovation. The principles that might have launched a great company or product 10 years ago are often seen as outdated today. But, do the principles of innovation really change that fast? Or can we continue to apply methods that worked for our fathers' generation?

Nearly 30 years ago, Peter Drucker, now known as "the dean of management philosophers," provided five concrete principles to guide innovation in his landmark book *Innovation and Entrepreneurship: Practice and Principles*.[1] It was one of the first attempts to clearly describe the value of innovation and to establish clear principles for practicing it. Drucker penned his principles at a time when Apple was inventing the Macintosh computer and Microsoft was just about to release the first version of the Windows operating system. Given that everything else we thought we knew about technology and innovation has changed, surely these "old" principles don't apply in today's competitive markets for cellphones, tablets, and web services? Or do they? Business, technology, and society may seem radically changed, but much of Drucker's catalog of effective practices remains vitally relevant to today's innovators.

1. Innovation is systematic and begins with an analysis of the opportunities that exist.

Drucker argued that an innovation must be matched against at least one of seven sources of opportunity in order to have a viable market. Drucker's opportunity sources include:

- Unexpected and poorly understood changes in the business environment
- Incongruities within a market arising from a mismatch between the efforts of established businesses and the needs of customers
- Unmet process needs in which the means of production, delivery, or service are not optimized
- Industry and market structural changes, most commonly created through rapid market growth or changes in government regulation
- Demographic changes that create new markets

- Social changes in perception about products and practices
- The emergence of new knowledge that enables new products or services

One of the biggest business, market, and government regulation shifts in recent time has been the introduction of the Affordable Care Act in the United States. This has certainly been a major shift in government regulation, but the changes in the health insurance market it has engendered arise from more than one of Drucker's opportunity areas. The ACA is an attempt to address demographic changes, as a major portion of the population moves into their retirement years. It expresses a change in social opinion, as more people believe that health care should be accessible to all citizens regardless of financial resources. It addresses the inefficiency of a healthcare system that has not been optimized to deliver affordable care to large numbers of people. And finally, it exposes and addresses the incongruities between market need and service providers.

This controversial government mandate is an expression of larger social and economic forces that have been at work for decades and are now triggering major changes. Those forces will not abate—they will only increase. All of this presents significant opportunity for innovation in products and services that are effective in this new environment. The changes to the health insurance business catalyzed both by the ACA and by the incongruities between business offerings and customer needs will lead to many new offerings based on ACA regulations and aligned with the business structures of healthcare providers.

But these all risk failure if they do not match the specific needs of the customers who are expected to use them. The new regulations increased the population of potential customers for health insurance, but they did not increase the capacity of the healthcare system. And what's needed is not necessarily just a scaling up of the current full-service healthcare system. New customers are more likely to create a demand for entry-level services and a continuing relationship with a provider who maintains personal history records. This suggests a need for walk-in care centers and primary care physicians as opposed to specialists. The challenge for innovators will lie in getting new services right to match demand without wasting money offering services that will not be used.

2. Innovation is both conceptual and perceptual.

New products can begin with an intellectual concept and be carried through design and prototyping based on a good idea. But they must be tested perceptually, through interaction with the user market. Without a perceptual understanding of the market, and of customers' needs and desires, there is a huge risk that a brilliant concept will fail to find a sufficiently large customer base.

The Apple iPad represents an astute alignment of technical concept and market perception. The iPad is an entirely new offering based on Apple's perception of a market for consumer entertainment and social services, as opposed to the traditional computer market focused on creating office documents and software. The iPad is ideal for streaming videos from YouTube, Netflix, or Amazon Instant Video. It is a convenient platform for reading digital books from Amazon, Barnes & Noble, or independent publishers. It is an effective music box, supporting iTunes, iHeartRadio, Rhapsody, Pandora, and Spotify. It is an efficient portal into social services like Facebook, Instagram, Twitter, and Pinterest. But it is a poor device for editing documents, building business presentations, or constructing spreadsheets. Apple understood the strengths and limitations of the new product, but perceived that the market for an entertainment computer was significant enough that the iPad could be a hit even without initial adoption by the business community. The company's technical concept and their market perception were aligned to identify and satisfy the needs of entertainment-oriented customers.

3. An effective innovation must be small and focused.

Customers and markets can absorb only a limited amount of change in their practices. Therefore, an effective innovation should be focused, easy to adopt into existing practices and small enough to be affordable. It should be focused on satisfying one need, not overturning the way customers perform many activities or replacing multiple devices all at the same time.

As first blush, it would appear that the modern smart phone violates this premise. But the smart phone is not a single, sweeping innovation, but a collection of small, focused ones. The cell phone was a simple product that performed a single function—mobile phone calls. Over time, these phones evolved a few additional programs, like text messaging and address books. With each iteration, the product grew in capability and complexity. Today's smart phones offer complete mobile computing platforms that serve as the foundation for hundreds of simple and focused innovations, in the form of apps.

4. Effective innovations start small.

Each new idea needs to be small enough that it can adapt to the changes it creates. Most innovations are "almost right," not quite finished. They must navigate multiple changes in design, functionality, and market positioning before they reach their technical and market sweet spot. Like a ship at sea, the larger the innovation is, the more energy and time it takes to change course. Adjusting the "almost right" product will require additional resources and time, all of which can be minimized if the size of the innovation itself is small. If the innovation is too large, then the changes required will be correspondingly large and could challenge the resources available to make the change and add significantly to the time required to fix a problem.

Google has embraced the idea of the small, "almost right" innovation from its beginnings. The company now generates a vast array of new products and services on a regular basis, most of them released in beta. This does not dissuade millions of users from snapping them up; in fact, these users expect Google's new products to be limited in capabilities and subject to constant updates. The company's now ubiquitous mail service, Gmail, was released in beta form in 2004. Potential users had to be recommended by another Google user even to get an account on the service, creating an aura of exclusivity that accelerated demand. Gmail remained a beta product until 2010—for six years, customers were expected to accept shifts in appearance and functionality, with features added and removed as a normal part of the beta development process. Google, on the other hand, was not obliged to retain earlier features or to make the service backwardly compatible with older versions, as most commercial software vendors are expected to do when they upgrade their core products. Starting small allowed the company the freedom to experiment without the constraints that come with a "finished" software product.

5. A successful innovation aims at leadership.

Being small is not the same as being mediocre. A successful innovation should aim at putting itself, and the company that produces it, in a leadership position that is distinctive enough to attract customers while creating barriers to competitors. A great idea that is poorly implemented will just show competitors where the market is and attract customers who will adopt a really excellent implementation offered by a following competitor.

The FitBit Force activity tracker, one of this Christmas's smash successes, is a clear example of the importance of leadership. The

FitBit Force was completely sold out in all physical and online retailers by the end of the year, largely because it offered the right combination of sensors, feedback, and app support to beat out competing products from established companies like Nike, Polar, and Jawbone. The design and functionality of the Force made it a clear leader in customer demand. However, the company has since become aware of issues with the materials used in the product's wristband, forcing a recall of all of the devices. FitBit Inc. is now experiencing the downside of being "almost right" with a product presented as final, and finding out whether they have the resources to survive the shift to the next generation of their product line.

Clearly, Drucker's ideas remain relevant to 21-century innovators, even 30 years after their first appearance. A management professor, consultant, or executive could construct an entire practice around these ideas and still make a valuable contribution to their students and clients. Perhaps the ideas in those old books aren't as faded as the yellowing paper makes them appear.

References

Peter Drucker, *Innovation and Entrepreneurship: Practice and Principles* (New York: Harper & Row, 1985).

Originally Published in *Research Technology Management*, July-Aug 2014

CHAPTER 6

A SPRING IN THE DESERT

"A library outranks any other one thing a community can do to benefit its people. It is a never failing spring in the desert."
- ANDREW CARNEGIE, 1903

Carnegie Library, Lamar, Colo

How do you stimulate education, creativity, and innovative thinking across an entire nation? In the 1880s, in a country just emerging from the destruction of the Civil War, one answer was to build a network of libraries to deliver the latest knowledge across great distances to a disconnected population. Much of that network was constructed with the largesse of Andrew Carnegie; from 1883 to 1929, Carnegie donated funds to construct 2,509 libraries around the world, 1,681 of them in the United States. By the end of the project, Carnegie Libraries formed over 50 percent of the library network of this country.

The functions of these libraries were similar to many of those we now access largely via the Internet. They were local hubs for knowledge collection, dissemination, and consumption. They tied the community together, both by filling needs for outside information and entertainment and by serving as both meeting places, for groups, and quiet sites of contemplation, for individuals—everyone came to the library at some point.

One of those nodes of knowledge was erected in my home town in 1908 and stood as the grandest building in the county for more than 60 years. As a 13-year-old boy, I found a particular kind of refuge there. Thirteen-year-old boys generally have few worries once released from the school day. They can look forward to cruising the town on a stylish stingray bicycle, followed by a little homework and maybe some television. But the trip from school to home was not always easy. As eighth grade began, I picked up a new tormentor. Marcus and his nameless sidekick seemed to appear from nowhere, demanding to know how much money I had and suggesting it would be best for me if I handed it over. This must have been an early step in Marcus's criminal career, because he had chosen a poor target. I rarely had any money at all; a good day would have yielded a dime or a quarter. The first day I escaped with a warning and a punch in the arm. But there was a second day, and a third, and a fourth. It became a regular pattern to be accosted for money and mildly beaten on my way home from school.

The library provided a solution to my Marcus problem. One day, instead of heading toward home after school, I made my way to the impressive steps and columns of the Carnegie Library. Marcus would be waiting two blocks further on, wondering what had happened to his easy mark. I climbed the steps and entered the library's cool, musty common area. I had only been there once or twice previously, so knew little about what lay inside. I browsed the shelves, looking for something to read. There may have been serious books on astronomy, physics, chemistry, history, or business. But I was 13; I gravitated toward a treasure trove of novels that were not available in the local drug store, the de facto literary hub in a town of only 9,000 people. I found the works of Edgar Rice Burroughs, Isaac Asimov, Arthur C. Clarke, and Robert Heinlein. These stories opened a world of adventure, heroism, and a fantastic future—a future set in a frame of solid, real science. While reading about space travel in a Clarke novel, I learned about the physics of rockets and the classification system for stars, and I gained a sense of the immensity of the universe. Burroughs presented heroes with high morals, integrity, and bravery in the face of danger. Nowhere else was the world colored with such promise and grandeur.

But as significant as these ideas were, what I most remember is the quiet. I had stumbled into a refuge from the commonplace chatter of a small town, which formed a different kind of network for circulating smaller, more self-centered ideas, usually less noble and less deeply considered. Within those walls, there were no interruptions; there were no friends eager for attention, no loudspeaker announcements, no sounds of construction, no demands for my attention. There was nothing to do but read in silence. Listen to the words. The library time became part of my daily routine—go to school, spend an hour in the library, then head out to more typical activities—providing much more than just shelter from a bully.

Carnegie libraries provided refuge and connection for generations of people, right up to the 21st century; even as radio and television competed for control of news and entertainment, libraries remained the repositories of knowledge and inspiration. Much of that changed with the Internet, which captured both a huge portion of that knowledge and almost all of the attention of a generation. So many writers have pointed to the Internet's usurpation of the roles once filled by television, radio, and printed books that it has become a cliché. But few have noted the demise of that old knowledge hub—the library, the place where contemplation and concentration were possible, where quiet study could connect a young reader to wide world of ideas. Today the World Wide Web, Wikipedia, YouTube, and digital books bring a flood of information to the eyes, ears, and fingertips of everyone who wants it, exactly when and where they want it. Much of that information is delivered in small, quick bites; web "surfers" dart from one little nugget of knowledge to the next, resisting the tedium of digging deeper. There is no longer a common physical place for quiet, uninterrupted reflection. Libraries remain special places, dedicated to storing and distributing knowledge, but often, no one is there to see them anymore. They are forgotten jewels, for some, relics of a past era.

In this connected age, it appears that innovative ideas are born of mass connectivity—via email, Twitter, Facebook, LinkedIn, Instagram, Snapchat, WhatsApp, and a host of other platforms. Learning and inspiration are now all about sharing. Connectivity has become the dominant lifestyle and workstyle; it is the medium through which idea generation and innovation are being conducted now and into the foreseeable future. Knowledge is no longer retrieved from a repository; it's generated in, and by, the network.

But, even in this age when no idea seems born of a single thinker, I cling to the habits that I learned in the quiet of that library decades ago. When it is time to explore a subject in depth or to write a meaningful article, I retreat to a quiet room and banish interruptions. I work diligently, but with patience. Deep knowledge and understanding take time to form from many pieces passing through my mind.

Mere access to information is not sufficient for deep thinking; it also requires focus and isolation of the mind. Great ideas do not emerge fully formed after scanning a few pages. Engaging with ideas, building knowledge, requires thinking through the same ideas multiple times, as well as enduring through periods of thinking nothing.

The place for this kind of thinking may never again be an elegant building fronted by inspiring columns and populated with paper books. When every book and document is accessible on a personal device, it becomes the responsibility of the reader to find or construct a refuge from interruption, a space for deep thought. The physical space may be a coffee shop, a park bench, or a home office—wherever it is, it must offer escape from the constant stream of electronic interruptions. When physical isolation was equal to mental isolation, libraries offered a quiet place to think. Now that information and interruption are delivered together, through the same device and often in the same application, thinkers must create new means of seclusion. Just as the problem is technological, there may be a partial technological solution—electronic devices already come with an "airplane mode"; perhaps they also need a "library" or "deep thinking" mode, a setting that allows access to information but blocks interruptions. But the responsibility now lies with each of us to create the isolation, the quiet, we need to absorb and sort information, acquire real knowledge, and create new ideas.

One warm day in 1975, I rode my bicycle to the grassy lawn of the Carnegie Library and watched as bulldozers destroyed the classic building to clear the land for a new city government complex. After standing on that spot for 68 years, the bricks, columns, masonry, and concrete steps crumbled and fell into heaps. Thank you, Marcus, for driving me into the safety of the library and the inspiration that it offered. Your bullying lasted only a few months, but the habits of reading, learning, and quiet thought have stayed with me for a lifetime. They have been a cornerstone of my lifestyle that I value and struggle to preserve in the information age. Without you, I might not have found this spring in the desert.

Where is your spring in the noisy desert today? How do you find the quiet you need to explore new ideas and think deeply?

Originally Published in *Research Technology Management,* Jan-Feb 2018

CHAPTER 7

ON THE GREEK ORIGINS OF INNOVATION
(καινοτομία)

"We shall be regarded with more affection by the Greeks, shall live in greater security, and be more glorious, and we may come to see our city secure and prosperous"

— XENOPHON on innovations for stimulating mining, tourism, and shipping in Athens, *Ways and Means*

"If children innovate in their games, they'll inevitably turn out to be quite different people from the previous generation; being different, they'll demand a different kind of life, and that will then make them want new institutions and laws."

— PLATO on protecting Greek culture, *Laws VII*

"Since men introduce innovations for reasons connected with their private lives [modes of living], an authority ought to be set up to exercise supervision over those whose activities are not in keeping with the interests of the constitution."

— ARISTOTLE on Greek political stability, *Politics V*

The current positive view of innovation is less than a century old. The leading figure in this transformation was Joseph Schumpeter, whose works evolved through the first half of the 20th century. It was he who exposed and solidified the modern view of innovation as a process of "creative destruction":

The opening up of new markets, foreign or domestic, and the organizational development from the craft shop and factory to such concerns as U.S. Steel illustrate the same process of industrial mutation—if I may use that biological term—that incessantly revolutionizes the economic structure from within, incessantly destroying the old one, incessantly creating a new one.

This process of Creative Destruction is the essential fact about capitalism. (Schumpeter 1942, p. 82)

Prior to Schumpeter's persuasive arguments, innovation was generally viewed as a corruption, inevitably and destructively disruptive to a well-designed society and business environment. Innovation introduced changes that threatened the livelihoods of innocent and productive citizens who had been practicing an established trade in an established form for generations. This negative view extended back to the early Greek origins of the term.

Historians trace the roots of the Greek term for innovation (καινοτομία) to the writings of Xenophon, who appears to introduce the term in a positive way, suggesting that some innovations could bring economic advancement for Athens. But in the works of Xenophon's more famous contemporaries, Plato and Aristotle, the term was used with a negative connotation; in their work, καινοτομία was fruitless change, deviation and manipulation in politics, culture, and entertainment for its own sake or to advance individual interests over those of society. For these philosophers, who were actively engaged in the thoughtful construction of an ideal society and sought to create a governing class guided by moral philosophy, that kind of change could only bring unwelcome upheaval. Their emphasis was on protecting that work in order to create a society that was both fair for all citizens and enduring through generations. They were very conscious of the real possibility that the Greek state could descend into a state of chaos and war under the influence of self-interested men. To prevent such catastrophe, Plato and Aristotle and others like them advised government control of almost everything and stood against any change, at any level of society, that was not government sanctioned, even to the play, toys, and music of children.

Both Plato and Aristotle cited the case of prominent and wealthy citizens who lost their wealth through bad business dealings or basic misfortune, for instance when their trade ships were lost in a storm, and then attempted to use their prominence to reduce or eliminate their losses by changing the laws of the country in their favor, perhaps by allowing them to access state funds to rebuild their wealth. To protect against this kind of "political innovation," Aristotle prescribed a number of principles that would seem to be rational even today:

- Avoid extremes in principles.
- Do not augment the power or honor of any one man out of proportion.
- Exceptional prosperity in one section of society is to be guarded against.
- Ensure that the number of those who wish the constitution to be maintained is greater than those who do not.
- Treat each other in a democratic spirit, that is to say, on an equal footing.
- Set up an authority for control.

Plato's and Aristotle's views prevailed for eons. After Xenophon, innovation was not described in a positive light again for nearly two hundred years, until Plutarch. Plutarch's *Lives*, a collection of biographies of great Greeks and Romans, admiringly described the great innovations in city government that were instituted by the Roman dictator Sulla, the innovative business enterprises of Themistocles, and the magnificent innovations in art by Stasicrates. But the negative and fearful view of the effects of innovation on a functioning society remained predominant in Greece, as they still do in many countries today. Even in the 21st century, it is the exception rather than the rule that countries not only allow but

Some Ancient Greeks on Innovation

Plato

"The overseers of our state must cleave and be watchful . . . against innovations in music and gymnastics counter to the established order." New songs are allowed but not "new way of song." (*Laws VII*)

"When the program of games is prescribed and secures that the same children always play the same games and delight in the same toys in the same way and under the same conditions, it allows the real and serious laws also to remain undisturbed; but when these games vary and suffer innovations . . . [children] have no fixed and acknowledged standard of propriety and impropriety." (*Laws VII*)

Children "hold in special honor he who is always innovating or introducing some novel device," but "the biggest menace that can ever afflict a state" is changing "quietly the character of the young by making them despise old things and value novelty." Change "except in something evil [or humorous amusements like comedy] is extremely dangerous." (*Laws VII*)

"The intermixture of States with States naturally results in a blending of characters of every kind, as strangers import among strangers innovations." Strangers are most welcomed, unless they bring in innovations in the city: magistrates "shall have a care lest any such strangers introduce any innovation." (*Laws XII*)

Aristotle

"All possible forms of organization have now been discovered. If another form of organization was really good it would have been discovered already." (*Politics II*)

"It very often happens that a considerable change in a country's customs takes place imperceptibly, each little change slipping by unnoticed." (*Politics V*)

"Even a small thing may cause changes. If, for example, people abandon some small feature of their constitution, next time they will with an easier mind tamper with some other and slightly more important feature, until in the end they tamper with the whole structure . . . . The whole set-up of the constitution [is] altered and it passed into the hands of the power-group that had started the process of innovation." (*Politics V*) champion innovation in government, business, culture, entertainment, and other facets of society.

Athens struggled for more than a century to create a stable and fair government that incorporated these principles; that struggle was still in progress when Plato and Aristotle were active. Today, the nations that allow innovation tend to be the politically stable ones. Correlation is, of course, not causation, but perhaps internal stability is a prerequisite for the ability to tolerate and benefit from innovation. Perhaps the acceptance of innovation comes from the kind of stability that gives a society resilience in the face of change. Where change has the potential to disrupt the entire social order, innovation is a threat to the well-being of all citizens. But where society is structured for resilience in the face of change, innovation can be embraced as a force that will benefit society as a whole, even though it may be destructive to specific pockets. This relationship argues for the need to create a social structure that is resilient to change from both the inside and the outside. Without that property, innovation can only be seen as a threat to be prohibited—a prohibition that, ironically, will lead only to further calcification and decay of the society. Maintenance of the status quo at extreme levels is actually the equivalent of the destruction of that status quo.

As Schumpeter observed, it would seem that democracy coupled with capitalism is one form of flexible stability that can tolerate and even encourage innovation as a means of improving the lives of citizens in the long run.

References

Schumpeter, J. 1942. *Capitalism, Socialism and Democracy* (Rpt. New York: Harper, 1975).

Originally Published in *Research Technology Management*, Jan-Feb 2019

PART II

INNOVATION INTELLIGENCE

CHAPTER 8

AI IS THE 21ST-CENTURY PRINTING PRESS

Much has been made of the contribution Gutenberg's printing press (1440) made to the spread of Martin Luther's *Theses* (1517), and hence to the launch and spread of the Protestant Reformation. Gutenberg's press enabled the distribution of thousands of copies, not only of Luther's work, but also of other religious documents and even entire Bibles, so that laypersons could access the documents of their faith and consider the roots of that faith privately, without the direct intervention of the clergy. A parallel version of this story can be found in medicine. Prior to the printing press, it was almost impossible to share widely the existing knowledge of human anatomy, physiology, and treatment; the most common way of sharing knowledge was through personal letters, which—by definition—were not widely shared and could not be

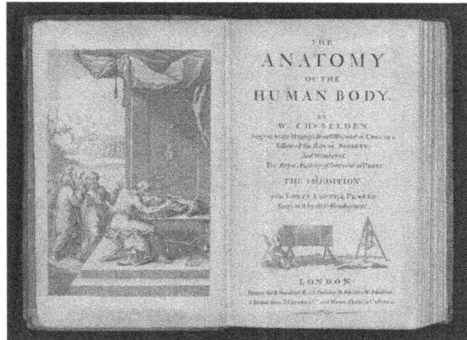

easily reproduced and distributed. The advancement of medicine was hampered—perhaps even stymied—by the lack of a medium for distributing the knowledge discovered by dispersed practitioners. Gutenberg's press created an efficient medium for distributing this rare knowledge to large numbers of dispersed practitioners.

Today, we're seeing the emergence of another technology that promises to have an equally powerful impact, not only in medicine but in a whole range of other fields—artificial intelligence. AI won't just reproduce knowledge to make it accessible to human learning; rather, it will gather and process information to create knowledge based on data streams too large and complex for humans to synthesize. But AI isn't going to change the world in quite the way you think. Artificial intelligence won't take over the world, at least not anytime soon, nor will it completely displace people. Rather, initially it will make people smarter and more efficient.

AI has been about to change the world for sixty years. It has been a promising academic field for the entire career of many computer scientists, and it has served as the basis of science fiction stories since before Isaac Asimov. But it has never really broken out in a big way. Its time is about to come. The Internet, cloud-based computing resources, unlimited storage media, the digitization of knowledge, and the opportunities to test and learn on all of the data in social networks, online video sites, and search results may be the missing pieces that have been needed for fifty years.

Consider this scenario: One day in the near future, an avid bicyclist takes a serious spill and ends up in the emergency room complaining of shoulder pain. The nurse asks him to stand and move about the room as she observes his posture and facial expressions. This initial assessment is used to get a sense of how serious the injuries might be. At the same time, a tiny camera captures the same information and delivers it to a cloud-based AI system for analysis and comparison with thousands or millions of similar cases. As the human nurse enters her opinions into a laptop computer, the AI returns its own analysis in an adjacent window. The nurse compares her own thoughts with those of the AI, making decisions about her next step as she merges the two data streams.

As the cyclist reclines on the bed, the nurse takes a verbal description of the events that led to his injury and the complaints he reports. The AI listens as well, converting the words into text and performing a global search matching combinations of events and complaints. The AI also analyzes the tone, inflection, and pacing of the cyclist's voice to estimate the level of stress the patient is experiencing, both physical and psychological.

The nurse peels off the cyclist's shirt to examine his shoulder. She notes the patterns of bruising and abrasion then looks at the shape of the swelling near the shoulder. She gently manipulates the arm, rotating the shoulder in different directions, and asks the patient to assess his level of pain as she explores the range of motion. Again, the AI watches and listens, comparing the information and its own automated assessments to an enormous database of similar situations.

Again, as the nurse enters her findings into the computer, the AI provides a 3D rendering of the shoulder that highlights probable areas of concern and rates the likely severity of injuries. Taking this data into consideration, the nurse sends the patient for X-rays, the orientations of which are partially guided by the AI's recommendations.

The X-ray machine is tied into the network so that the AI receives the digital images and correlates them with previous observations of the patient, both its own and the nurse's. Processing the images in multiple dimensions, the AI compares these to any previously stored images of the patient's shoulder, as well as to healthy and injured images of millions of other shoulders. This analysis is performed in the same time it takes a human doctor to examine the new images on a computer screen, but in that time, the AI has processed the information through 100 different algorithms from multiple angles. It applies the intense scrutiny of thousands of computer processors, hundreds of algorithms, and thousands of previous cases to this patient's case in the time it takes for one human doctor to apply his or her own personal experience and training.

Everything the computer AI is learning is being combined with the lifetime medical records of the patient to provide recommendations for his treatment and to equip future medical practitioners with a picture of the patient's lifetime experience. De-identified versions of the data are also being cataloged in enormous databases to improve the performance of AIs on similar cases. Thus, this one case will improve the experience of thousands or millions of patients and capabilities of thousands of AIs, doctors, and nurses treating similar injuries.

Together, the human clinicians and the AI adjutant determine that the cyclist has suffered a level 1 separation of his AC joint. The bruising and abrasions make it look much worse than it is, and the patient's pain is largely due to external injuries rather than internal damage. The prescription is for anti-inflammatory medications, range-of-motion exercises, antibiotic cream with bandages, and a follow-up appointment with an orthopedist. These instructions are entered into the computer and processed by the AI. A prescription is automatically sent to the patient's regular pharmacy, as well as an order for the cream and bandages to be picked up with the prescription. The AI also recommends a conveniently located and well-reviewed orthopedist near the patient's place of work. Accessing that practice's scheduling calendar, it offers two appointment times and books the one selected by the patient.

In this case, the patient is treated by a traditional, well-educated and experienced medical team, but that team is augmented by the knowledge and experience of thousands of other physicians and patients, in the form of the data accessed by the AI. The AI's input guides decisions and improves the quality of care, as well as making the entire experience more efficient for both the clinicians and the patient.

The technologies needed to realize this scenario are in existence today; IBM's Watson is already working with physicians in much the way I've described above. But these systems have not been widely integrated or made a routine part of the medical delivery system. And healthcare AI is just the tip of a huge iceberg, one hinted at by Apple's Siri and Microsoft's Cortana, as well as others that are less widely publicized or commonly used. These are the first indicator of a coming future, one in which customers rely on the expertise of a single local information source or service provider, but that expertise is augmented by the experience of thousands or millions of others, captured in the cloud, and the computing power of a fully functional AI.

Medicine is already seeing the beginnings of the changes intelligent augmentation will wrought, but many other fields are poised for a similarly revolutionary change. We have already become accustomed to semi-intelligent telephone customer assistance. Wall Street investment firms have been using AI algorithms to manage trading strategies for over a decade. Surveillance and intelligence collection systems acquire far more data than a human team can evaluate, and AIs are being created to analyze that data more efficiently. Air traffic management systems track and coordinate thousands of flights simultaneously.

Others are on the cusp of change, driven by cultural and technological pressures. Our aging population will require intelligent in-home assistance, a friend who is always observant and available; AIs can provide steady vigilance as well as data collection to more carefully track medical status and, when needed, a direct line to alert human caretakers. Nuclear disasters, which create conditions too dangerous for humans to remediate, can be contained by intelligent machinery operating in concert with human engineers. Children's toys and computer games can benefit from more realistic playmates and flexible algorithms that provide tailored imaginative—and educational—experience. Business executives would welcome an assistant that constantly monitors, analyzes, and integrates every piece of news relevant to their market, as well as their own personal data streams, equipping them with a much more complete and deeply considered picture of potential opportunities and coming challenges.

Of course, these changes don't come without costs. AIs could improve national security, but at a cost to personal privacy; reduce air traffic accidents, but with a loss of human understanding of the system; improve care for the elderly, but reduce their real human contact; fascinate children and gamers, again with a reduction in human contact; and reduce business missteps, but without requiring human leaders to acquire a complete understanding of the misstep that might help them act more effectively in the future.

AI and machine learning are being touted as one of the—if not *the*—next big things of the 21st century. Their success in financial markets showed how speed and repetition can turn small profits into huge fortunes. Their success with self-driving cars is prompting a reconsideration of legal and ethical frameworks around automation and driving itself. The opportunity to supplement professional decision making in medical clinics and executive suites is too attractive and valuable to ignore. All of these benefits, and many more that can as yet only be imagined, make AI augmentation an attractive—indeed, almost unavoidable—target for investment, research, and business adoption.

To some, the advent of AI represents the beginning of an era of menacing computer intelligence, the kind represented by *2001*'s Hal and *Terminator*'s Skynet, one that will culminate in the demise of the human race (see "The Disturbing Side of AI," below). However, the fears that we imagine in a few years do not control our decisions as much as the opportunities that we perceive in a few quarters. As a population, we will grapple with the impacts of AI just as we did with industrial automation and digital computation. Initially, large groups will feel threatened because they perceive automation of some part of their jobs as demeaning to them and their human limitations. Many will grapple with the real effects of Schumpeter's "creative destruction," as their hard-earned skills, career achievements, livelihood, and positions in society become obsolete.

But as the real capabilities and limitations of the technology become clear, its advantages will be appreciated and adopted by many, even as it disempowers and displaces others. Ultimately, humans will redirect their efforts and reshape their society around AI. The generation that follows will build on this foundation, just as the children of factory saboteurs learned to work with industrial machines.

Originally Published in *Research Technology Management*, Jan-Feb 2017

CHAPTER 9

THE INTELLIGENT DECISION

I t's 1880—should the son of a Pennsylvania farmer plan to take over the family farm, or move to Pittsburgh for a job in the new steel mills? It's 1980—should a Pittsburgh high school graduate pursue a union job in the steel mills or a college degree in electronics? It's 2020—should a college graduate pursue an MBA or an advanced degree in computational AI?

At each of these times, we were on the cusp of a radical shift in the technologies driving new industries. In each case, companies and individuals had to make significant decisions about their future based on incomplete and inconclusive information. Today, we are in a similar place—wondering whether this AI renaissance will be the one that drives a revolution as significant as the industrial and information revolutions before it—or this wave of enthusiasm will recede as previous ones have. The field was born with the AI Spring of 1956, when key developments made machine intelligence seem just around the corner, and enjoyed a long season of optimism until the harsh AI Winter of 1974. With each subsequent AI spring, scientists emerge from their labs with new techniques that could perform impressive tasks, but fall short of delivering a generally intelligent computing machine. After successful applications generate excitement and also map the limitations of the approach, the new techniques become generally understood—and are promptly stripped of their AI label. They are judged to be clever, perhaps even ingenious, but not intelligent, and the chill winds of the next AI winter blow in.

The field seems defined by a promise of intelligent behavior that can never quite be achieved. That apparently unattainable promise represents a siren's call, luring the next generation of scientists further out into the seas of computation.

The newest incarnation of AI is built on deep learning applied to huge volumes of Internet traffic and shopping data, powered by vast cloud computing resources that can tackle translation and reasoning tasks that were beyond the reach of the computers available a decade ago. The astonishing power of these systems has led many to suggest that we are about to experience a second machine age, one in which AI will challenge knowledge workers for their jobs. This "knowledge Taylorism," some say, will reshape the workplace, moving jobs that require limited reasoning into the hands of machines and making them partners with the humans who remain to process more difficult tasks (Brynjolfsson and McAfee 2016). In the first industrial revolution, humans found themselves working alongside machines on the factory floor. In this second machine age, humans will work alongside machines everywhere in the organization, creating a more productive, precise, and profitable business enterprise than either could deliver alone.

We now recognize that intelligent machines do not have to possess the general intelligence that humans possess to become working partners with us. These machines and algorithms can start small and work their way toward more responsibility. AI can begin as a simple tool embedded in other programs, such as the autocomplete function in search engines and word processors. From there, AI applications can take over the text chatbot services found on many web pages, using their cloud computational resources to process the sentences provided by customers and provide accurate answers using human sentence structures. Taken a step further, AI may operate as a peer to humans, perhaps processing simple insurance claims independently while allowing humans to focus on complicated cases—and consequently reducing the number of humans necessary to do the work. And finally, AI might become a manager of humans, for instance by listening to human customer service agents and suggesting changes in their phraseology, speaking pace, and tone in order to align the human's behavior with profiles known to be successful in dealing with customers. Here, the AI will direct the work of many humans, while at the same time documenting and reporting their progress in following the AI's instructions. Each of these examples describes a real system that is in limited use today and will potentially impact huge numbers of jobs in the near future (Malone 2018).

We have seen this kind of job shifting before, when the Internet spread around the world and call center jobs were moved from the

United States to India and China, displacing tens of thousands of jobs. If the predictions about AI are correct, the next shift will not be to a different country, but into a different dimension, from the physical world to the digital world, where nationality and locality are fungible. This fact has caught the attention of national governments. The United States, China, India, and others all recognize that even very low national wages cannot compete with an AI employee who draws no salary or benefits, needs no office or parking lot, asks for no raise or promotion, and is available 24 hours a day, seven days a week, surviving purely on electricity and silicon.

These countries are equally terrified and ambitious to become the home of the next generation of intelligent workers. Without knowing exactly how to accomplish this goal, China has initiated a national AI policy; the United States belatedly followed (Lee 2018). When large, powerful nations focus their vast resources on a growth plan, it usually portends immense government spending with immediate opportunities for businesses, universities, and individuals. Our 2020 student considering graduate school will find full-ride scholarships to the best colleges, universities will find research and endowment money for new departments and degrees programs, and businesses will find ample government contracts for prototypes and new products.

But even heavy national investments cannot contain the new AI citizens within geographic borders. Once created, AI workers will be able to travel unhindered wherever they can be employed most productively and most efficiently. The only competitive advantage will be for the country that can invent these citizens and license their use around the world.

References

Brynjolfsson, E., and McAfee, A. 2016. *The Second Machine Age: Work, Progress, and Prosperity in a Time of Brilliant Technologies.* New York: W. W. Norton & Company.

Lee, K. 2018. *AI Superpowers: China, Silicon Valley, and the New World Order.* New York: Houghton-Mifflin, Harcourt.

Malone, T. 2018. *Superminds: The Surprising Power of People and Computers Thinking Together.* New York: Little, Brown & Company.

Originally Published in *Research Technology Management,* May-June 2019

CHAPTER 10

A "MACHINE LEARNING FIRST" COMPANY

ave you noticed how much more accurate Google search results have become in the last year? ... Machine learning.

Have you seen the improvements when you search for pictures and videos online? ... Machine learning.

Have you been surprised by the predictive word and phrase suggestions in Gmail? ... Machine learning.

Have you experienced improved language translation from Google Translate? … Machine learning.

Google and a few others are creating and playing an entirely different game than everyone else.

All these improvements were the result of years of effort and experimentation in developing and perfecting machine learning models and software. Hundreds of computer programmers have been shifted from writing new search and translation algorithms, to writing machine learning models capable of performing search and translation better than humans can. The improvement in performance is significant, as is the reduction in the amount or code and effort required to customize the models. Jeff Dean of Google has reported that 500 lines of machine learning code in TensorFlow has replaced 500,000 lines of traditional code in Google Translate. With the right machine learning model, the system performs better, requires less code, and is significantly easier for human staff to learn, master, and maintain.

At Google, machine learning is gradually changing how all the company's products and services work. Josh Cogan, another senior Google engineer, has stated that the company is currently using hundreds of different machine learning models across all its products, services, and internal operations. The company's leaders have declared Google a "machine learning first company." This statement means that when creating or improving a software product, Google looks first for a way to do it with machine learning as opposed to hand coding it the traditional way. And Google is not the only company making big investments in machine learning. Amazon, Microsoft, IBM, and Netflix are all making similar moves to evolve their products and services and build an advantage over competitors who have not yet grasped the power of machine learning.

What makes machine learning so powerful is its ability to examine millions of data points, find patterns, and make decisions that adapt its performance in real time. Human business analysts can't work as broadly or as quickly as machine learning algorithms can. The company that masters these tools essentially doubles, triples, or even quadruples its workforce at a fraction of the cost of adding human employees.

This trend in cognitive labor mirrors the transformation of physical labor during the industrial revolution, when machines replaced large numbers of manual laborers, reducing the time to create products while also improving their quality. Industrialization eliminated some kinds of jobs, but it also created new ones; someone had to manufacture, monitor, and maintain the machines. The role of the human shifted from making products to making (and fixing) the machines that make products. This shift is occurring now for workers who make software. They will become coders, monitors, and maintainers of the machine learning models that will provide the software's capability. Millions of coders who know how to program in the traditional manner will have to learn to program or monitor machine learning models; many other workers who now spend their days analyzing data will need to learn to analyze and measure the performance of machine learning models, instead of working directly with data.

That change is coming—it's even under way—but it's not here yet, at least not on a wide scale. Business media would have us believe everyone is already turning machine learning into improved business performance. But actually, only a few are. It is still very early days; the power of machine learning has been applied only to a few, quite similar classes of problems. Only the rare one-tenth of one percent of companies are on the frontiers, unraveling their businesses, experimenting with machine learning, and finding ways to remake their businesses with these tools.

Discovering how to apply machine learning more broadly remains a challenge. It cannot be applied to all business IT problems just yet. The databases and transaction processes associated with sales and order placement, for instance, have yet to be subjected to machine learning. Machine learning algorithms can examine images, audio, video, streams of transactions, medical images, radio signals, and other similar forms to identify patterns that are not explicitly obvious by other methods; those patterns can then be mapped to actions to be taken—that's how the software replaces humans. But it does not work when the problem is unique or too unusual, when there are too few previous instances to generate a pattern.

In a 1993 NPR interview, science fiction writer William Gibson said, "The future is already here; it's just not very evenly distributed." Google and its peers are on the front edge of the future with respect to machine learning. It might be time for the rest of us to begin learning how to play this new machine learning game. The machine learning revolution will change the way we write software, process data, respond to customers, and design products. As a result, it will increase the velocity of business and the velocity of change within a business, or a market, or an industry. This future has already happened at Google. When will it happen for your company?

Originally Published in *Research Technology Management*, Jan-Feb 2020

PART III

INNOVATION THEORY

CHAPTER 11

NULLIUS IN VERBA—
THE BIRTH OF INNOVATION

O n the night of October 9, 1604, Italian citizens recorded their astonishment that the universe had changed for the first time in centuries. As they looked to the night sky, they found a new star had appeared. They knew nothing about the process for the creation of stars; rather, they believed that the heavens were fix and immutable. The same stars in the same arrangements made up the map of the heavens for their entire lives and going back for generations. What did it mean to man and society if the heavens could change their composition?

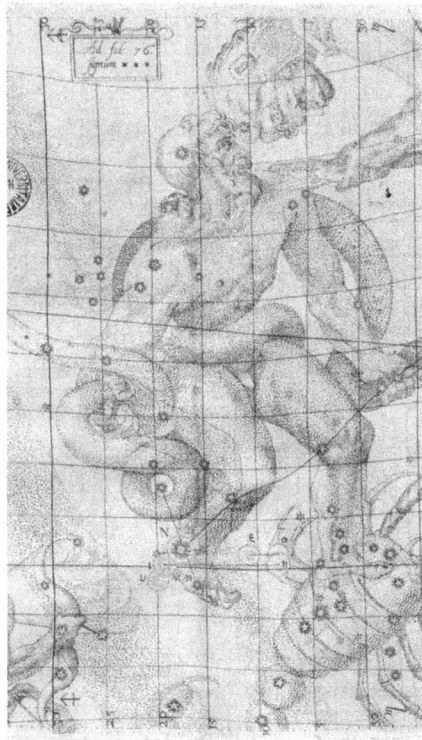

Johannes Kepler was especially intrigued by this new object. He studied this new star in detail, writing an entire book on his observations, *De Stella Nova in Pede Serpentarii (On the New Star in Ophiuchus's Foot)*. His book contained an original drawing of the position of the stars with the shadow of Ophiuchus framing the constellation. The new star appears in the heel of the constellation, labeled 'N'. As a result of this work, SN1604 is now also known as Kepler's Supernova.

The supernova and Kepler's methodical observations of it planted the seeds for a new approach to scientific discovery. The appearance of a new star opened people's eyes to new possibilities: If there is more in the heavens than has been known in the past, then perhaps there is more on earth as well. If even the heavens are subject to change, questioning the accepted knowledge received from the ancients might be permissible, beneficial, and even essential. SN1604 did not change the physical earth, but it did begin to change the mental framework of earth's inhabitants.

This natural philosophy took some time to coalesce. Its first expression might be seen in the formation of a small group of natural philosophers in the area around London, England in the late seventeenth century. On November 28, 1660, this "Committee of 12" met to hear a lecture by Christopher Wren and afterward formally declared themselves "a Colledge for the Promoting of Physico-Mathematicall Experimentall Learning." Two years later, they were officially recognized by the Crown with a Royal Charter and the loose title "The Royal Society of London for Improving Natural Knowledge"—and the motto "Nullius in verba," Latin for "Take nobody's word for it" (Royal Society 2012).

Christopher Wren and the Committee of 12 laid the groundwork for the acceptance of inquiry, innovation, invention, and change in the natural and mechanical worlds. They made it acceptable to question what is and extended their thinking to the question of what might be. For Wren and his compatriots, knowledge and wisdom were not—as they had been believed to be—derived exclusively from the "wisdom of the ancients." They offered a new structure for pursuing knowledge, a new understanding of innovation as a positive force.

However, the boundaries of scientific inquiry were not yet defined. Astrology and astronomy were one practice, and alchemy was just beginning to grow into chemistry. Scientists explored fortune telling and divination alongside mathematics and physics. Gradually, through the development of an objective method of inquiry and the accumulation of results from repeated, well-structured experiments, the parameters of science and a disciplined method of inquiry emerged.

As natural philosophers uncovered how the physical world worked, they provided essential knowledge for manipulating that world to meet practical needs. The birth of real science led to the birth of real engineering. This mother-daughter pair equipped thousands of intelligent and inquisitive minds to invent, discover, and innovate their way out of the Dark Ages, creating a scientific Renaissance that would eventually give us the Industrial Revolution. The work of the Committee of 12 and their generations of successors opened people's minds to the possibilities of inquiry, enabling them to question accepted knowledge and test their own theories about how the world might work and how it might be changed for the better. Nullius in verba.

Aristotelian tradition maintained that air was a basic element. This was accepted as fact from 350 BC until the late eighteenth century, when Joseph Priestley discovered oxygen and several additional gases in the mixture formerly classified simply as "air." Nullius in verba.

Newtonian physics sufficiently described the behavior of the natural universe for over 200 years. But Albert Einstein changed all of this with his proposition of the general theory of relativity in 1916, changing our understanding of the entire universe for a second time.

Today we see yet another transformation from our growing understanding of quantum physics. Nullius in verba.

Everyone knew that computers were specialized tools for large corporate and scientific jobs. Only dedicated scientists were interested in or capable of using them. The whole world agreed with DEC Founder Ken Olson's famous 1977 declaration, "There is no reason for any individual to have a computer in his home." But Steve Jobs, Steve Wozniak, Bill Gates, and Paul Allen created a personal computer appliance, which the entire world eagerly brought into their homes and allowed to reshape their lives. Nullius in verba.

"Nullius in verba" is one of the most essential ingredients for innovation and one of the most difficult to support and maintain in a large organization. For an individual working alone, it may be the only path to creating something notable. But for an established organization, the idea threatens the foundation that has made the organization successful and that serves as the structuring force among its people and departments. At some point in the evolution of countries, religions, and businesses, the value of maintaining what has been achieved in the past seems to surpass the potential value even of a true breakthrough.

In the four centuries since Kepler, scientists have struggled with these two competing forces—the need to respect and apply existing knowledge and the need to question everything, even the wisdom of centuries. Isaac Newton believed that he was not discovering or inventing calculus, but was rather reintroducing lost knowledge that had been known in ancient times (Dolnick 2011). As Thomas Kuhn posited in his influential book, *The Structure of Scientific Revolutions* (1996), new ideas do not supplant established ideas, until new scientists supplant their forbears. Established leaders give weight to the trove of accepted knowledge through their positions of authority. Innovative businesses can wait for "the Kuhn Force" to change the hierarchy of leadership or they can take intentional steps to free new thinkers from the constraints of past successes.

Can we practice "Nullius in verba" today as it was originally intended nearly four centuries ago? Do we still have a naiveté that will allow us to venture into new territory or to return to old questions with new eyes? Can we seriously ask, "What if?" What if a core belief is not true? Where might a "ridiculous" experiment really go? What if we take the risk to explore truly new territory?

References

Dolnick, E. 2011. *The Clockwork Universe: Isaac Newton, the Royal Society, and the Birth of the Modern World.* New York, NY: Harper.

Kuhn, T. S. 1996. *The Structure of Scientific Revolutions.* Chicago, IL: University of Chicago Press.

The Royal Society. 2012. History of the Royal Society. *The Royal Society.* http://royalsociety.org/about-us/history/ (accessed February 27, 2012).

Originally Published in *Research Technology Management*, May-June, 2012

CHAPTER 12

BUREAUCRACY AS INNOVATION

Sumerian table written in cuneiform (ca 3100 BCE) describing the distribution of beer to citizens. Source Wikipedia: https://commons.wikimedia.org/wiki/ File:Early_writing_tablet_recording_the_allocation_of_beer.jpg

A ncient Sumer (ca 3500 BCE), like most of the world at that time, faced a number of challenges in operating a kingdom and maintaining a functional society. One key function essential to the survival of the people was planting, raising, harvesting, and distributing food crops. People had learned much earlier what made crops like wheat and barley grow, but distributing the harvest remained a significant challenge. Every community had evolved its own methods for accomplishing this, most of them far from efficient. In order to prevent starvation, kingdoms needed some technology for optimizing the growth and distribution of food.

What was needed was a means to capture information about the volume of crops and make decisions about how to distribute food. The technology that partially solved this problem was writing, in cuneiform script, usually on clay tablets. Now, the decisions were clear; writing captured them for execution. But another innovation was needed to distribute the information and ensure the appropriate actions were taken. The answer, the rulers of Sumeria found, was a system of representatives who could apply and enforce standards across the kingdom—an administrative bureaucracy. Sumerian rulers began to use government administrators to improve and standardize the distribution of food during the Uruk period (4100–2900 BCE). The bureaucratic application of standards led to fewer deaths from starvation, longer life spans, and a healthier life over all.

This structure for managing basic societal necessities was so successful that rulers began to apply it to other areas. Gradually, as an understanding of the benefits of this new innovation spread, administrative bureaucracy spread through China and Eurasia. The countries that adopted an administrative bureaucracy prospered and grew. In the most successful systems, bureaucrats received specialized training and became part of a professional hierarchy within which they could advance to higher positions over the course of a career.

These civil servants often replaced older systems based on favoritism and corruption. Thomas Meadows, a 19th-century British consul to China, attributed that empire's long survival to "the good government which consists in the advancement of men of talent and merit only" (Meadows 1847).

But bureaucracy has a down side. Today, we're all familiar with the ways in which it interferes with individual productivity and stifles creativity. Even as it spreads known best practices, encouraging systemic productivity, bureaucracy stymies the creation of the next generation of better practices by limiting the dissemination of new, untested methods. Lacking a means to evaluate the large number of competing new ideas, it tends to suppress them all.

This has long been the case. In the 12th century, Italian bankers and merchants used the Roman numeral system for accounting and calculation. Roman numerals were difficult to manage, requiring a special table abacus to perform addition and subtraction. As a result, a unique tradecraft emerged for those who could master this skill; these educated practitioners became an essential part of all banking, commerce, and government services. But, in the 1228 edition of *The Book of Calculations*, the mathematician Fibonacci (born Leonardo Bonacci) introduced Arabic numerals to European scholars. This new number system, which was much easier to deal with, quickly spread to merchant and banking leaders. They converted from Roman to Arabic numerals in droves, rapidly rendering the table abacus obsolete. Fearing the loss of their livelihoods, the abacus tradesmen petitioned city governments to stop the use of Arabic numerals, and the bureaucracy, seeing only the immediate negative impacts of the change, responded—by siding with the established practices and outlawing the use of the new, disruptive—and, in their eyes destructive—number system. Then, as now, bureaucracy defended the entrenched positions and attempted to stop the adoption of new and better methods, which in the long run would benefit and enrich all of society.

This, then, is the central contradiction of bureaucracy: it is a central innovation that may serve to suppress innovation. Sociologist Max Weber, who was among the first to describe the power of bureaucracy, believed it to be the most rational means of exercising control over human beings (Swedberg 2005). He saw that the material destiny of the masses of society was tied to the continuous functioning of the bureaucratic apparatus. Without it, he argued, the core protections and support that define a government and a society would lose their efficiency (Jacoby 1973).

But bureaucracy can be, and frequently is, misapplied. The precision, stability, and uniformity it provides are essential for maintaining an army, enforcing the law, operating public transportation, providing reliable water and sanitation, administering consistent taxation, and enabling sustainable growth and expansion. But its success in these essential areas leads to its application to all endeavors, without regard for the size or purpose of the project. When its rigid rules are imposed out of context, it can create as many problems as it solves— or even stop a good project in its tracks.

Bureaucracy is necessary to support progress, but at the same time it is uninspiring. Its rigid processes and standards snuff the spark of imagination, lack consideration for the human condition, and do not tolerate change. John Stuart Mill (1861) believed that "bureaucracy stifles the mind and tends to become a pedantocracy." Weber warned that bureaucracy leads to a "polar night of icy darkness" in which the rationalization of human life traps individuals in a soulless cage of rule-based, rational control (Ritzer 2009). Robert Merton (1957) described it as leading to "trained incapacity" and "over conformity."

Eight centuries after the advent of Arabic numerals in the West— Italian merchants simply ignored the prohibition—we still struggle to find a way to balance the plusses and minuses of this very effective form of organization. It is not clear what the final outcome of this struggle will be for our society, as even the older societies of Europe and Asia have not arrived at the final stage of bureaucratic control. It would seem that over-bureaucratization would lead either to social and economic stagnation or to a revolt against government control. An alternative in which bureaucracy is limited in an effort to foster and support innovation would appear to be desirable, though it's not clear that such a balance is possible—even though bureaucracies themselves recognize innovation as an activity needing protection, as evidenced by functions such as patent systems. These attempt to apply bureaucratic and legal processes to protect new ideas so that their creators can reap the benefits and remain encouraged to innovate. This protection addresses the tension between providing for the public good and protecting the rights of an individual.

Corporate bureaucracies have also struggled with this balance. Some have tried to isolate innovation from the general bureaucracy, to allow it to grow unimpeded. The establishment of separate R&D departments in dedicated facilities created islands that are free from some bureaucratic control, but without triggering dissent in departments that don't have similar exemptions. More recently, companies as diverse as Lockheed and Apple have created internal "skunk works" operations to achieve similar goals. These experiments suggest that separation and compartmentalization of work functions in ways that allow the organization to tailor levels of bureaucratic control to each function's unique needs for levels of bureaucratic control is essential for realizing the benefits of both bureaucracy and innovation.

From this perspective, it's easy to see that bureaucracy is a tool like any other, one that must be applied properly to provide a benefit. The success of bureaucracy in one function does not necessarily recommend its full application to all functions. Finding the balance requires acknowledging that fact and finding ways to limit the reach of bureaucracy where it is not conducive to productivity.

References

Jacoby, H. 1973. *The Bureaucratization of the World*. Berkeley: University of California Press.

Meadows, T. 1847. *Desultory Notes on the Government and People of China*. London: Wm H. Allen. https://archive.org/details/desultorynoteson00mead

Merton, R. K. 1957. Bureaucratic structure and personality. In *Social Theory and Social Structure*, 195–206. Glencoe, IL; Free Press. http://www.sociosite.net/topics/texts/merton_bureaucratic_structure.php

Mill, J. S. 1861. VI—Of the Infirmities and Dangers to which Representative Government is Liable. In *Considerations on Representative Government*. http://www.gutenberg.org/ebooks/5669?msg=welcome_stranger#2HCH0006

Ritzer, G. 2009. *Enchanting a Disenchanted World: Revolutionizing the Means of Consumption*. 3rd ed. Thousand Oaks, CA: Pine Forge Press–SAGE.

Swedberg, R. 2005. Budgetary management. In *The Max Weber Dictionary: Key Words and Central Concepts*, 18–21. Stanford, CA: Stanford University Press.

Originally Published in *Research Technology Management*, Jan-Feb 2016

CHAPTER 13

MODELING DISRUPTION:
A METHODICAL RESPONSE TO THE NEXT HOT TECHNOLOGY TREND

"I think that concludes the agenda for our strategic initiatives," the CEO said, wrapping up the strategy meeting. "Are there any new items before we adjourn?"

"Sir, I would like to offer some thoughts on a new technology," John said. John was one of the newest executives on the team. Very intelligent and quick thinking, the CEO thought.

"Yes, John, please go ahead."

"Many of you will have noticed the extreme energy and press coverage around the new offerings for cloudchain technology. I believe there are several places where it could improve our own internal operations and increase efficiencies. I would like to suggest that we create a task force to look at cloudchain in more detail and determine if we should be pursuing it for our business operations."

The CEO nodded. "Thank you for the quick summary, John. We are all aware of the current interest in this new technology. We have also been approached by a number of consulting firms offering to help us with a strategy study on how it will impact our business. This seems to be the season for the next big idea. These things roll through the business community about every two years and everyone seems to think they have to move on the idea as fast as possible or be left behind. Your cloudchain falls squarely into that category."

"But sir, if it really is as transformative as everyone says, then our market could be disrupted and our customers siphoned off to competitors or new startups," John argued.

"That might be true," the CEO said. "*If* it is as transformative as everyone says. But we can never know how true those claims will prove to be. John, you are one of our best and brightest new executives. How would you propose to determine the real disruptive power of this cloudchain?"

"Well, we might start by hiring one of the leading consultancies," John offered.

"Yes, that would be the lazy way and the expensive way. A more effective and personalized approach would be to designate you to study the problem in the context of our own industry and our own customers," the CEO countered. "In fact, that is your assignment. I want you to go forth, master this technology, understand its power in our industry, and report back on what you find. You can use internal resources, but you cannot hire outside help."

"Yes sir," John replied, both nervous and eager at this new charge.

"Well then, the meeting is adjourned." The CEO was pleased with John's eagerness and with the change of orientation in pursuing a solution. As everyone filed out of the room he called, "John, another moment please."

"Yes, sir."

"Understanding how a new technology will impact an industry, existing customers, or even the world economy does not come from mastering all the details of how that technology works. You will understand it well enough with a few days of study and conversations. What you need is a set of models for how our systems already work so you can explore how a new technology might disrupt those models. I want you to create several different models of the pieces of our business that would be affected by this cloudchain idea. Models focused on customers, internal operations and processes, market orientation, and our partnerships. Looking at models will force you to examine how all of the bits of information you have interact with each other. Hopefully this will reveal poor assumptions, biases, and inconsistencies in your thinking."

John disappeared into the labyrinthine offices of the company and was rarely seen for weeks. Rumors told of his meetings with scientists in the R&D labs, managers on the front lines with customers, and veterans who had been with the company for three or four decades. They reported the incessant questions John asked and the diagrams he sketched in an oversized book. Occasionally, John's boss received emails or calls from other managers, asking if they should give so much information to the young man. The answer was always "Yes, certainly. He's working for the CEO on this project." At the mention of the CEO, everyone became an eager supporter of John's project, offering whatever he needed. Some even agreed to participate in an informal working group for a fraction of their time.

A month later, John requested a meeting with the CEO.

"So what did you learn about this new technology's impact on our business?" the CEO asked.

John responded, "I have been really impressed with the experience of our leadership team across the company. They all told stories of dealing with something similar in the past. One old guy even talked about how putting telephones in every office improved internal efficiency, connections to suppliers, and relationships with customers."

"I'm guessing that was Joshua? He actually worked with the founders in the early days," the CEO responded.

"Right, he mentioned that," John said. "So, with each hot new technology, from the telephone, to the Internet, to web-based business forms, the leadership seemed to go through similar excitement and gyrations. Eventually they realized they needed to create a model of how something new would impact them. And, as you said, this always led to multiple models because they just did not understand the entire market or country well enough to make one big model."

"And what were the people you talked to able to contribute?"

"They all had references to past models, work done by consulting firms, and a number of influential business books that were based on models. Clayton Christensen's books *The Innovator's Dilemma* and *Seeing What's Next* came up several times. Also, the work of Eric von Hippel, James Utterback, Everett Rogers, even Peter Drucker. Joshua talked about something called Quality Circles, which I had never heard of. But the consistent message was that heavy thinkers have been creating and prescribing models for a hundred years. It reminded me that we had learned about several of these in business school. But somehow referring back to them when we have a real disruptive innovation was not my first response. The excitement of something new moved me to take some kind of action as quickly as possible. It seemed that we could lose our position if we didn't act fast. I wanted to *do* something, not think about some academic ideas."

The CEO grinned, "I see it all the time. We need your energy and my experience working together to run this company. How many models do you have?"

"Four that are central to my thinking; a couple more had interesting details. First, a customer impact model. If we move some of our services to a cloudchain solution, how will customers be impacted? Will they experience something better than what we do for them now? And, just as important, will this new technology allow us to create customers from people who do not use our services at all right now?

"Second, an internal operations model. Can the technology improve internal operations in ways that do not necessarily touch the customer but which would make us more efficient or reduce internal errors?

"I made cost reduction a separate model. Since we can measure cost improvements in dollars, that model is much more straightforward and quantitative, whereas internal operations are measured in time, steps, and error counts, which are trickier to convert into dollars.

"My fourth model is focused on our suppliers and external vendors. I wanted to see if the technology could change those relationships, either by reducing the costs or maybe even by eliminating the need for some of the providers.

"By this point I felt that I had pretty much covered the waterfront. I spent a little time working on a marketing model—Is it possible for us to promote the technology as an advantage of working with us? But in the end, most of this work blended into the customer model, so I put it over there."

The CEO said, "That is quite a mouthful of work. Do you think you're finished?"

"Well, I don't think we need additional models to help us study the impacts of the technology, whether it is cloudchain today or something else tomorrow. But the details and accuracy of all the models can always be improved. Changes added by myself or the little team I have created could make minor improvements, but not necessarily enough to move the needle on decisions we might make with them. I think any significant improvements would come from people who haven't seen the models before. Outsiders will understand them from a fresh perspective and may be able to point to missing pieces that have totally escaped us," John said.

"That is very insightful," the CEO said, impressed. "It's time to let some fresh eyes look at these models and give you feedback on what is missing. Are you ready to go on the briefing circuit? I can arrange half a dozen meetings right away."

"Yes sir. And I want to express my appreciation for this assignment. I have learned a lot more about handling new ideas than I expected. I understand how cloudchain is just one in a constant stream of new ideas that senior leadership has to deal with every year. We can't chase them all, or even consider them without some kind of reference framework like these models."

Cloud computing, machine learning, virtual reality, Segway scooters, satellite cellphones, wearable computers, blockchain—all of these are or were popular new concepts that promised to change the world, overturn established business practices, displace industry leaders, shift success into the hands of competitors who adopted them. All of them, analysts argued, had to be adopted by incumbent firms if they wanted to have any hope of surviving. But science, engineering, and the media create world-busting technologies faster than they can all be studied and absorbed by any company. What should executives do in the face of this constant onslaught of new tools and revolutionary technologies? How to sort out the truly useful from the merely overhyped, and then how to develop a reaction? The approach offered here is based on the professed practices of Charlie

Munger, the legendary co-leader of Berkshire Hathaway, who advises people to create multiple mental models to guide their thinking about investments, business decisions, and life directions. This same approach is the basis of multiple business books that offer a model for understanding the impacts of new innovations, including the works of Clayton Christensen, Henry Chesbrough, Eric von Hippel, James Utterback, Everett Rogers, and Peter Drucker. All of these have offered models as tools for understanding the impacts of innovation. And like the innovations they model, these books tend to become all the rage when they are first published and then quickly fall off the radar as tools for real business strategy.

Originally Published in *Research Technology Management,* July-August 2018

CHAPTER 14

THE EVOLUTION OF INNOVATION

In 1856, William Henry Perkin was an 18 year-old student at Britain's Royal College of Chemistry. He was working toward an anti-malarial drug that was important to the British Empire as it expanded into Africa. But he stumbled onto a coal-tar derivative that was particularly effective at staining silk material into a rich shade of purple. At a time when dull, earthen colors had dominated clothing for two centuries, Perkin realized that a vibrant and stable purple dye was a very valuable product. He quit the university against the protests of his professors and established a factory for producing the dye. His father invested the family's entire fortune in the endeavor and his brother quit a job in the building trades to manage the new business. By 1857, Perkin's factory was producing "Tyrian Purple" for sale to commercial silk dyers and he was working on new dyes for wool and cotton (Buderli, 2000).

The success of Tyrian Purple as a commercial venture led chemists across Europe to focus on this market in the hope of making their own fortunes. Over the next fifty years major companies like Bayer, Hochst, BASF, and AGFA built their fortunes on the creation of new dyes. The sustained demand for dyes built new factories, created a demand for educated chemists, raised the importance of a university education, and provided employment for thousands of workers. The plot of this 19th century story is closely matches that of current Silicon

Valley computer, software, and web services companies—the curiosity, hard work, and good luck of one person leads to the creation of one unique product, followed by the invigoration of an industry.

FORMAL STUDY OF INNOVATION IS NEW

The creation of the aniline purple dye was an invention. The application of that dye as a commercial product was an innovation. Both invention and innovation are very old processes. The history of weaponry, machinery, and transportation are all filled with instances of invention and innovation that transformed individuals, companies, countries, and economies. But, as old as these practices are, the formal study of innovation is relatively new tracing it roots back to the works of Burns and Stalker in 1961 and Rogers in 1962. In *The Management of Innovation*, Burns and Stalker clearly separated mechanistic from organic environments. In a mechanistic environment it is best to create standard processes, rules, and hierarchy to improve the efficiency of the organization. But organic environments require a different approach, one which recognizes the importance of unique skills and knowledge, as well as the means to stimulate these toward solving new problems and creating new products. Organic working environments require employees to use their own knowledge and judgment to solve a continually changing set of problems. In *Diffusion of Innovation*, Rogers investigated the means by which new ideas propagate through a society. He was most interested in the social factors that allowed ideas to prosper and identified five variables that determine the rate of adoption of a new idea or product: the attributes of the innovation itself, the type of decision required to adopt an innovation, the communication channels through which the idea is carried, the nature of the social system into which the idea is introduced, and the extent of a specific change agent's promotional efforts. These ideas, published 45 years ago, were some of the first models of innovation.

It could also be argued that the ideas of Fredrick Winslow Taylor in 1911 and Joseph Schumpeter in 1942 were about innovation. Taylor may have invented new shapes and sizes of shovels to implement his ideas about labor productivity, but the real innovation was in recognizing that traditional methods of factory work were inefficient and could be improved if a scientific mind were allowed to enforce practices on the working hands of laborers. This idea so threatened the positions of laborers that Taylor found himself defending it in court and before Congress. The immediate affront to labor was much clearer to most people than the long-term economic benefits of efficiency, profitability, and lower costs. Schumpeter and David Wells pointed out that inefficiencies in business cannot be sustained indefinitely (Perelman, 1995). New ideas and new technology will transform a single practice or an entire industry to eliminate these inefficiencies. Schumpeter's "creative destruction" is the march of invention, innovation, and change across the face of society and business. Though it is painful to many people, it accrues to the good of the entire society. Individual interests cannot stop this march, but they can rush to get in front of it so that they are on the creative side of change, not just its destructive side.

> Could the invention of the methods of innovation be the greatest invention of the 20th century?

TRANSIENT VALUE OF INNOVATION

Today we describe innovation as an activity or action that creates value from materials, processes, or ideas that are available to many people, but which have not been recognized or applied by others. But, like Perkin's discovery of purple dye, the success of an innovation will draw others seeking to capitalize on similar ideas or seeking to copy them outright. This value is a very transient thing that, unprotected, will flow from an inventor to a competitor without regard to claims of ownership. In fact, the entire system of patents, copyrights, and legal IP protection exists to allow inventors and innovators to prosper from their work. We recognize the value of innovation to society, the economy, and business and are eager to foster the personal and organizational investments that are required to make it happen. However, the time limits on these legal protections must also balance the good of the creator with the good of society. Legal systems allow a temporary monopoly as a means of encouraging and rewarding innovation, but prevent long-term control of an idea that can benefit an entire society.

Once the value of innovation was both recognized and protected, it became desirable to analyze the practice and attempt to formalize it for repeatability. Establishing a business on a single innovation is a great entrepreneurial venture. But sustaining an ongoing business through random and haphazard innovation is much too risky. I suggest that our understanding of innovation has also evolved so that it can be practiced in a disciplined, organized, and directed manner; or one can continue to pursue it in a random and haphazard way (Figure 2.1).

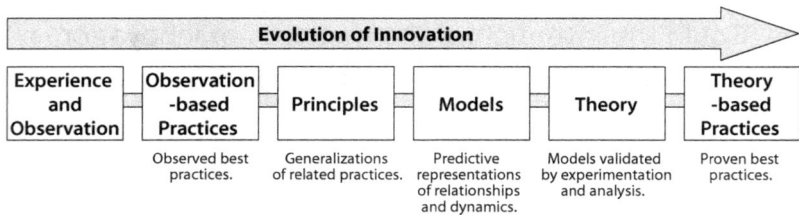

Figure 2.1. Innovation has evolved from individual experiences and observations to practices based on theories.

HOW INNOVATION EVOLVES

In its native form innovation begins with *observation and experience*. Someone notices something valuable and repeats the activity to repeat the benefits. This may come from a new technology, a new process, or a unique application of something that has been around for years.

These observations lead to *practices* that appear to capture the value noticed in observation. The practice-stage of innovation focuses on selecting and fine-tuning specific practices to improve results. However, practices are very limited and not immediately extensible to other businesses, activities, or products. Therefore, practices are soon extended into principles.

Principles are rules that seem to generalize the important aspects of specific practices. The foundational literature of management, the works of Henri Fayol (1916) and Chester Barnard (1938), focused on extracting general principles from their years of experiences, observations, and practices. They hoped to provide a foundation upon which big businesses could be built and thrive, without reliance on the intuition and experience of a single outstanding leader.

Where principles derive from historical information, *models* attempt to structure this knowledge so that it can be extrapolated to future applications. Models attempt to describe both facts and the relationships between them to create a dynamic representation that can identify what the past and present mean to the future. Most models do not claim to be exact representations of a real system, but rather capture useful information and make predictions better than a less structured view of that information.

Innovation *theories* attempt to get at an absolute truth about a system. They separate legend, intuition, precedent, and varied practices from objective truths that can be counted upon to deliver exact results. Theories are based on experimentation and analysis. Because of their exactness, the breadth of theories may be limited by what can actually be proven. But theories also serve as a solid foundation upon which new ideas and experiments can be built.

Finally, *theory-based practices* close the loop. These replace limited observations with limited theories in describing the most appropriate practices. Theory-based practice is the core foundation of science. But in management, the entire spectrum of innovation from Figure 1 is at work daily across the world economy.

MOVING TO THEORY-BASED PRACTICE

Historically, practicing businesses have worked from observation, experience, practices, and principles, while models and theories have been the realm of academics, researchers, and consultants. Combining these two communities allows practitioners to move from observation-based practices to theory-based practices. Since this progression of understanding takes time, it assumes that the derived theories remain applicable to the environment from which the data was collected. In a fast changing field, it may be impossible to create theories at a pace that remains current with the environment in which it will be applied. In some fields, change is so rapid that practices based in observation are better than those based on theory because of the currency of observations and the age of the information upon which a theory is built.

The sheer number of popular books on innovation indicates that most practices and models are based on observations rather than theories. A search of the Amazon.com database identifies 14,353 available titles on "innovation". When the search is narrowed to business titles published since 1988, the list shrinks to 3,451. Figure 2.2 shows how these titles are distributed over each year, clearly indicating a rapid increase in 1997—the year that Clayton Christensen's *Innovator's Dilemma* appeared.

Innovation Books Published Annually

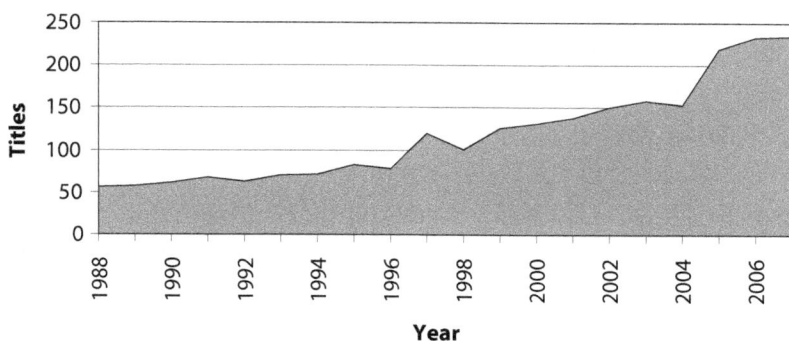

Figure 2.2. *Annual book titles on the topic of innovation has risen from 56 titles in 1988 to 279 titles in 2011, with a peak of 324 in 2010. This results in a cumulative twenty-three year library of 3,451 titles. Source: Amazon.com*

Buderi's account of 19th century Bayer identifies one of the company founders as a chemist. The company showed its persistence in creating new products using every means possible. It initially tried to create new products on the manufacturing floor as an integrated part of the production process. When this failed, they experimented with hiring chemists and allowing them to remain at their universities to interact with other faculty members, seeking to isolate their chemists from the distractions of production problems. This did not yield results within one year, so the company pulled the chemists back to the factories and assigned them to specific production lines. However, Carl Duisberg, one of the young chemists, settled into a research lab and spent the next two years inventing three new colors for the company. Bayer's persistence in trying new methods along with the arrival of a talented researcher led to new products, higher profits, and a much larger research organization centered around Duisberg, who later became a member of the company's board of directors.

THE INNOVATION IMPERATIVE

Regarding invention, innovation, change, and renewal, Charles Kettering told the United States Chamber of Commerce in 1929, "I am not pleading with you to make changes. I am telling you you have got to make them—not because I say so, but because old Father Time will take care of you if you don't change. Advancing waves of other people's progress sweep over the unchanging man and wash him out. Consequently, you need to organize a department of systematic change-making." (Buderi, 2000)

Alfred North Whitehead famously said that, "The greatest invention of the nineteenth century was the invention of the method of invention." (1925) What has been the greatest invention of the twentieth century and how are we using it in the twenty-first century? Could it be the invention of the methods of innovation? What do we really understand about innovation? That will be the primary focus of this column. We will explore current principles, models, and theories of innovation and make some attempt to understand how best to use them.

Definitions

Innovation:
- the introduction of something new (Merriam-Webster Online)
- an idea, practice, or object that is perceived as new by an individual or other unit of adoption (Rogers, 1962)
- an activity or action that creates value from materials, processes, or ideas that are available to many people, but which have not been recognized or explored by others

Model:
- to abstract from reality a description of a dynamic system (Fishwick, 1995)
- a representation of an actual system (Banks, 1998)

References

Buderi, R. (2000). *Engines of Tomorrow: How the world's best companies are using their research labs to win the future.* New York: Simon & Schuster.

Burns, T. and Stalker, G.M. (1961). *Management of Innovation.* Oxford: Oxford University Press.

Rogers, E. (1962). *Diffusion of Innovations.* New York: Simon & Schuster.

Taylor, F.W. (1911). *Principles of Scientific Management.* New York: Harper & Brothers.

Schumpeter, J.A. (1942). *Capitalism, Socialism, and Democracy.* New York: Harper & Brothers.

Perelman, M. (Summer 1995). Retrospectives: Schumpeter, David Wells, and Creative Destruction. *Journal of Economic Perspectives*, 9(3), 189-197.

Fayol, H. (1916). *General and industrial management.* London: Pittman Publishing.

Barnard, C. (1938). *The Functions of the executive.* Cambridge, MA: Harvard University Press.

Christensen, C. (1997). *The Innovator's Dilemma: When new technologies cause great firms to fail.* Boston, MA: Harvard Business School Press.

Whitehead, A.N. (1925). *Science and the modern world.* New York: Mentor Books.

Fishwick, P. (1995). *Simulation Model Design and Execution: Building digital worlds.* Englewood Cliffs, NJ: Principle Hall.

Banks, J., ed. (1998). *Handbook of Simulation: Principles, methodology, advances, applications, and practice.* New York: John Wiley & Sons.

Originally Published in Research Technology Management, May-June 2008

CHAPTER 15

PARDON THE CONSTANT INTERRUPTION

Job Opening
Senior Research Scientist

Major research facility is seeking a world-class scientist to lead several innovative projects in advanced power generation. Successful candidate should possess the following qualifications and capabilities:

- PhD in scientific field from a leading university
- History of patent development and scientific publication
- Familiar with state-of-the-art in power generation and relevant scientific publications
- Experience leading teams of diverse professionals
- Ability to think deeply about complex problems
- Persistence to remain focused for long periods of time
- Responsive to management queries via e-mail, telephone, and text messages at all hours
- Active Facebook network of other scientists (at least 200 scientific friends)
- Active Twitter feed with daily or hourly replies to other scientists
- Current LinkedIn profile with more than 500 connections
- Accessible 24/7 via electronic forums with response time under 30 minutes to all queries

Candidates should submit CV via e-mail, send Twitter message to HR, and contact director of laboratory via LinkedIn. Unsuccessful candidates will be notified within 30 minutes. Prospective hires will be interviewed via Skype during the next three days, with a decision made this Friday.

P rior to the introduction of the telegraph system, information moved from one place to another at the speed of a steam locomotive, a dashing pony express, or a lumbering ocean liner. Letters, newspapers, books, and scientific papers required months to create and distribute. With this paucity of new information, the advantage went to those who had the mental ability to use information in interesting and unique ways. In a world in which information moved slowly, the competition was not the person who had more information—everyone had more or less the same information—but the person who could do more with the information he or she held.

The telegraph opened the door for rapid delivery of textual information. The fastest Pony Express took days to carry a message across the country; a telegraph operator could send the same message in minutes. The telegraph gave birth to the Information Revolution, but it is now the pony express modern communication. The speed record for sending Morse Code was set in 1942 by Harry Turner who succeeded in sending a message at an average rate of 40 words per minute. Even assuming that it was possible to maintain this constant speed for a 24-hour period, this would amount to only 57,600 words in a day, about 140 printed pages. How many thousands of pages of information does the Internet publish in any given day? How much information can you access in just a few seconds from your smartphone?

The frantic pace of information dissemination has real advantages. The person who knows what to do with information has easier access to a wider field of knowledge than ever before. We have tools and sources that provide information both passively, waiting for a user to go and find it, and actively, pushing it into users' hands with an announcement of its arrival. We can cast a wide network in search of new information, bringing more information and a more diverse set of opinions to every question.

But there are very real disadvantages, as well. Most of us were programmed to assume that a ringing telephone signifies something important, requiring a response; too often, we apply that same mental model to the constant flood of e-mails, text messages, tweets, social network updates, and RSS feeds. Our lives too easily become a constant round of frantic activity whose only aim is to keep abreast of an unending stream of "important" interruptions. Attending to these interruptions becomes a full-time job, the sole object of our attention, with the (apparently unreachable) goal of the empty inbox—or multiple inboxes on different information services.

This frantic routine leaves no time for inner reflection or the long, slow consideration of difficult problems. And with so little time to consume the flow, we are able to respond to it with only brief and quickly composed answers. We no longer create long treatises on a single topic, but publish dozens or hundreds of short, shallow answers daily on a wide variety of topics. With this condensation we all begin to think like consumers of information, rather than as its creators. We strive to create only the information that an audience will remember later; the sound-bite—or, perhaps, "memory-bite" becomes the medium of choice. The summary or the witty rejoinder becomes the work itself.

This is hardly an environment that encourages the sort of deep thinking that produces great works of art, insightful solutions to complex problems, or breakthrough new technologies. Nobel-quality problems cannot be solved in 15-minute chunks wedged between electronic interruptions.

THE GREAT ESCAPE

What we all need, especially research scientists and those grappling with large, complex problems, is an uninterrupted space of time, thought, and energy; time to reflect and to take action on the new information flooding in. It is not the digital connection, interruption, or contribution that creates a better product, but the reflection and deep thinking that follow it (Powers 2010). In this reflection, new ideas blossom. But without it, the entire process is an uninterrupted stream of interruptions.

How do we balance the demands of the always-on, always-connected information age with our own need to reflect and regroup? Is it possible to hold the expectations of bosses, coworkers, and professional peers at bay long enough to reflect on incoming information? How can we create a respected space to think deeply?

A number of authors have offered suggestions for mapping out a space for reflection.

Make disconnection time. The most basic method is simply to refuse interruption. Establish daily intervals during which you turn off electronic networking tools and focus on the work at hand. Balance the amount of time and energy devoted to connecting with others, with the time and energy spent focused on larger projects and questions. For example, some people come to work at 8:00, but do not check their e-mail or cellphone until 9:00. That first hour of the day is dedicated to uninterrupted productive work. Others have created a protocol in which they check e-mail only at specified times during the day. One professional described to Powers (2010) his practice of checking e-mail at 11:00, before lunch, and again at 4:00, prior to finishing for the day, condensing all of his e-mail–based interruptions into just two hours per day, leaving at least six hours for him to focus on other tasks.

Filter connections. You can manage incoming information, so that only the most important messages create an interruption. In most e-mail programs, you can create filters that route messages from important people in your life to one folder, and everything else to another.

This allows you to be prompt in responding to family members, the boss, an important client, or specific team members. But it allows you to ignore messages from the rest of the world until you are ready to deal with them. A similar strategy can be applied to the routing of cellphone calls. Calls from some numbers trigger the ringer while others are shuffled immediately to voicemail, which you can manage according to your own priorities rather than responding to the caller's sense of urgency. Twitter and RSS feeds can be similarly managed using applications available for computers and smartphones.

Take recovery breaks. Finding a reflective space may mean more than simply escaping distraction. It also means finding a mental space in which reflection can happen. Often, this can be achieved simply by escaping the interruption routine. Loehr and Schwartz (2003) recommend taking a recovery break at least every 90 minutes. Get away from your workspace for 10 minutes. Stretch your body and mind, eat a healthy snack, drink water, and perform a physical activity. Break the obsessive lock that can develop between people and their inboxes, alerts, and lists of action items. Periodically reenergizing ensures longer daily performance and more sustained concentration.

Escape transaction processing. In computer services, "transaction machines" are computers that exist simply to handle incoming queries, process them, deliver an answer, and return to the queue to get another query. Humans are not transaction machines, although the daily round of e-mail, twitter, and text messages can make us feel that way. Unlike transaction machines, humans have the ability to see the bigger picture, to understand the value of actions that are being taken, and to choose to work on the bigger picture. Stand back from the constant stream of queries or your list of action items. Think about what the organization or you personally are trying to accomplish. Identify the best approach to achieving that. Then select actions that contribute to that larger goal.

DEPTH VS. BREADTH

There was a time when depth of thought was driven by the scarcity of information and connectivity. Einstein was able to sit and think about his theories while working in the Patent Office because his only source of stimulation was printed journals. He relied on his own ability to perform mental experiments. He had to work through problems for hours because there was no ready source of stimulus that he could turn to for inspiration or information.

Today we are blessed and cursed with the ability to reach out and touch almost every piece of information and every person instantaneously. We have the opportunity to spend our entire lives connecting and collecting. In addition to this we have created a culture that expects everyone to be instantly connectable, instantly interruptible. Each of us now has the responsibility to manage both our depth of thought and our breadth of thought. Some jobs require 100 percent breadth of thought and action. Others require 100 percent depth. But these extremes are rare. Most of us occupy a middle ground where we are expected to mix both, but are left to our own devices to discover the best balance. Our external society pushes for higher levels of connection at the expense of internal reflection. But many of us feel the internal yearning to stop connecting and spend more time developing really rich ideas or products. We just need the courage of conviction and the understanding of our leaders to dedicate sufficient time and energy to the reflective parts of our jobs.

References

Powers, W. 2010. *Hamlet's Blackberry: A Practical Philosophy for Building a Good Life in the Digital Age.* New York, NY: HarperCollins Publishers.

Loehr, J., and Schwartz, T. 2003. *The Power of Full Engagement: Managing Energy, Not Time, Is the Key to High Performance and Personal Renewal.* New York, NY: Free Press.

Originally Published in *Research Technology Management*, Jan-Feb 2012

CHAPTER 16

PRODUCTIVE FACTORY TOWNS

Productivity isn't everything, but in the long run it is almost everything. A country's ability to improve its standard of living over time depends almost entirely on its ability to raise its output per worker. World War II veterans came home to an economy that doubled its productivity over the next 25 years; as a result, they found themselves achieving living standards their parents had never imagined. Vietnam veterans came home to an economy that raised its productivity less than 10 percent in 15 years; as a result, they found themselves living no better—and in many cases worse—than their parents.

— Krugman 1992, 9

L onghua, China, is a very new city that expresses a very old idea. A suburb of the city of Shenzhen, Longhua is home to 300,000 people, most of whom work for the same company. In fact, the infrastructure, basic services, and governmental support for the town are all provided by that company. Longhua is a modern "factory town," created and operated by Foxconn Technology Group. Foxconn, which manufactures high-end electronics like the Apple iPhone®, the Sony Playstation®, and Dell computers, built Longhua from the ground up as a way to consolidate workers and factories in the same location, improving productivity and increasing efficiency. Longhua is one reason the company was able to create manufacturing facilities that produce millions of units of sophisticated electronics at costs far below any Western factory.

Foxconn operates Longhua for the same reasons that companies have created factory towns for centuries: the commercial demand for products can grow faster than the ability of existing towns to increase their population, housing, and services to meet the demand more naturally. Chinese "factory girls" come to Longhua to earn money for their families and escape rural poverty just as men and women left their family farms for better-paying factory jobs during the Industrial Revolution. In the United Kingdom, "mill towns," built and managed by textile manufacturers, sprang up in Lancashire, Yorkshire, and Cheshire in the 1800s. At the beginning of the 1900s, similar towns sprang up all across Massachusetts, Connecticut, and the northeastern United States. Hershey, Pennsylvania, was built by Milton Hershey to house workers for his fledgling chocolate company.

The Silicon Valley area, with its concentrated need for a unique mixture of labor, might have started as a company town, but quickly evolved into a unique commercial ecosystem which included a local government that was able to provide the necessary services. The international demands for the products from Silicon Valley quickly drew in other high-tech companies and extended the eco-system from its original core into adjacent and complimentary industries that could leverage the unique talent mix that had been created. The area transcended the traditional company or factory town of previous centuries, and became a model that other regions and other countries strove to imitate. But even this advanced eco-system demonstrated that concentrated labor is more productive labor.

In the century since the last heyday of mill towns and mining towns, our Western image of work has evolved. We now see work as something separate from our personal lives. We intentionally locate our homes away from our jobs—often many miles away. We bridge this separation with a daily commute into the office and back home at the end of the day.

But the separation of work and home is an expensive luxury. Its costs are measured in hours of unproductive driving, barrels of wasted oil, miles of expensive highways, and armies of government employees to maintain and protect the infrastructure. None of this contributes to the productivity of the people, the companies, or the governments that support it.

Asian and Indian economies do not have this same wealth. Separation of home and work is a luxury they cannot afford. In the West, we have become suburbanites, but in Asia, the factory workers who assemble the tools of our high-tech lives—smartphones and televisions and videogame consoles—live in urban factory towns. The ability to create these towns provides emerging-market manufacturers a competitive advantage over Western producers. The attraction of factory jobs and the availability of factory towns make it possible for companies to create just-in-time communities to meet fast-growing demand for products like the iPhone.

Seen in this light, the factory town is not a remnant of the Industrial Age, but a competitive tool for concentrating labor, increasing productivity, and fostering collaboration. Information technology may allow us to concentrate knowledge and skills in a new way, but it does not eliminate the need for this concentration. Commuter cities cannot compete with efficient factory towns in productivity and cost. These structures repeat throughout history because they are so powerful. If we don't want to relinquish the freedom we have to live where we like, we must find a way to capture the productivity of the factory town and eliminate the waste, in energy and in human resources, represented by the suburban commuter.

Because there are many different kinds of work today, there can be many different kinds of factory towns. As Chinese companies have shown, assembly-line manufacturers may still rely on more traditional factory towns (with perhaps some modifications to acknowledge modern developments in workplace safety and human rights), but companies whose products require high degrees of internal collaboration and communication will seek different structures. In knowledge-based work, telecommuting can eliminate the waste of transporting workers every day while still allowing a virtual concentration of labor. Some companies have added to standard telecommuting systems the practice of "local podding," in which employees are assigned to physical office buildings based on their home locations rather than their reporting chains. So an office is no longer dedicated to a specific function, but is instead designed to accommodate the skills of those who live in its vicinity. This connects people to corporate resources, ensures that they maintain contact with corporate values, and allows them to use the company network to reach distributed coworkers. Telecommuting, podding, and similar practices improve productivity by restructuring the way work is done, without forcing people to move onto a specific company campus, or into a factory town.

Information technology may allow us to create virtual factory towns for some kinds of work, concentrating labor without concentrating bodies. Changes like these are essential if every country does not want to return to company towns like Longhua, China, or Hershey, Pennsylvania.

References

Krugman, P. 1992. *The Age of Diminished Expectations: US Economic Policy in the 1980s.* Cambridge, MA: MIT Press.

Originally Published in *Research Technology Management*, Mar-Apr 2011

PART IV

INNOVATION PRACTICE

CHAPTER 17

MAJOR LEAGUE INNOVATION

The American national game of baseball. Grand match for the championship at the Elysian Fields, Hoboken, N.J. / lith. of Currier & Ives.
http://www.loc.gov/pictures/item/90708565/

"**M**anagers tend to pick a strategy that is least likely to fail rather than picking a strategy that is most efficient. The pain of looking bad is worse than the gain of making the best move." This sounds like guidance from a management consultant trying to initiate changes within a major corporation. But it is actually a quote from

The Hidden Game of Baseball, written in 1960 and published in 1984, in which Pete Palmer expresses his understanding of the way professional baseball teams were created, players were recruited, and games were played. Palmer was part of a small underworld of curious minds who were analyzing baseball statistics and arriving at conclusions about how the game worked that were very different from the principles upheld by Major League Baseball (MLB) managers, owners, and players. Amateur sports statisticians, digging through the records in a quest for a higher understanding of the game, created a whole new set of ideas about how a team should be built and how baseball should be played. But they had no connection to or influence over Major League Baseball itself, so their ideas remained a minor sideshow for more than 20 years—until Billy Beane, the general manager of the Oakland As, one of the league's most unsuccessful teams, hired Paul DePodesta as a "sabermetrician," applied data analysis to his organization, and began winning baseball games with one of the lowest-paid teams in the league. This transformation represented a major disruptive innovation—it demonstrated that the data analysts really had discovered a recipe for turning low-cost players into winning teams with only minor changes to coaching and other structures. Today, this story is famous thanks to Michael Lewis's bestselling book *Moneyball* and the movie that followed.

Over time, as Beane repeated his success in multiple years, cracks began to form in the traditional thinking about how to create a winning team and a profitable MLB business. Teams from Toronto to Boston began to imitate Oakland's methods and to hire away the management talent that had implemented it.

Hidden knowledge. Disruptive ideas. Fringe subcultures. Entrenched power players. Legend taken as fact. All of these conspired to prevented new ideas from displacing the historical wisdom driving the business of baseball for more than two decades. But as long as everyone followed the same set of established rules, there was no need for anyone to worry about the impact of new ideas. That worked until just one team escaped the accepted wisdom and adopted a new set of rules. And Billy Beane kept working at those new rules, perfecting the approach until his team was consistently beating much "better," richer teams.

Definitions

Billy Beane has continued to implement his approach with the As since the 2002 Moneyball story, with much success. The model has spread throughout the league, changing the way teams with limited resources buy players and structure their bench. But it has had only a minor impact on the biggest, richest teams in the league. The relative standings, for instance, of the New York Yankees and the Oakland As is roughly equivalent to what it was in 2002, even as the salary disparity between the two teams has continued to widen. The As payroll has increased by just 14 percent while the Yankees has increased by 65 percent—and the win-loss records of the two teams have remained roughly equal (Table 1).

If salaries in professional baseball do not translate into better win-loss ratios, could the same be true in our industrial R&D departments? Does a hidden hiring model exist in our industries as well?

Table 1.—Comparison of salary and win-loss records,
Oakland As vs. New York Yankees

2002	Wins	Losses	Games Behind	Payroll
Oakland As	103	59	0	$41,942,665
NY Yankees	103	58	0	$125,928,583
2014	**Wins**	**Losses**	**Games Behind**	**Payroll**
Oakland As	88	74	10	$47,799,500
NY Yankees	84	78	12	$208,830,659

The As story proves that a new idea is powerless as long as no one will adopt it. But when an upstart company decides to stream movies on the Internet, an impoverished country establishes manufacturing facilities that are 90 percent less expensive than those available elsewhere in the world, a café abandons food to focus solely on coffee, or a baseball team hires players based on base hits rather than spectacular home runs, then the competitive field changes. To the established way of thinking, it appears that a revolution has appeared overnight and no one could have seen it coming. But in truth, powerful ideas take decades to develop a following. Everyone can see it coming, but no one takes it seriously because it is not part of the historical lore of the industry. Not until one competitor uses the idea to turn the tables on its competition do other companies scramble to understand what has happened, trying their best to compensate, often with the goal of reestablishing the playing field of the past and returning to the rules they understand.

But once the idea is unleashed, the historical balance is gone, and it will never return. It has created a new normal that must be understood and mastered, at least until the next crazy idea comes along and works. The world evolves in nature, in society, and in business. The seeds of that evolution usually grow at the fringes where they are unleveraged, but almost always recognized and ignored.

Professional sports are portrayed as precisely planned enterprises in which the talents of all members of the team are greater than the sum of the individual parts. But this stereotype is not completely accurate. In truth, teams are dominated by superstars and designed by gut instinct, past experience, and tradition. The story of the Oakland As demonstrates the extreme opportunities for improvement that lie unexplored in a business that is highly competitive, potentially profitable, and extremely talent driven. This is an environment in which competition is broadcast to millions of viewers every week and the stakes for losing are both financial and reputational. How likely is it that similar opportunities lay buried in businesses that operate largely in private with little exposure to the eyes of outsiders?

Most companies are much more opaque than professional sports, and their competitions play out over years rather than hours. The impact of change is difficult to measure and losing scores can be hidden by time and accounting. Are you alert to the real, productive metrics of your department? Are you aware of the crazy ideas that exist at the fringe of your business? Are you ready for the landscape to shift under your feet? Or are you playing the game differently and about the shift the landscape under your competitors' feet?

Originally Published in *Research Technology Management*, July-August 2015

CHAPTER 18

THE MODERN ART OF INNOVATION

"Good business is the best art."
— Andy Warhol, 1975

A single-serving bottle of Coca-Cola is available from any convenience store for about 1 dollar. The ingredients inside and the bottle itself cost about 2.5 cents. In other words, combining the ingredients with a brand name and an image creates a 4,000 percent price markup on a simple product that can be produced by anyone. Globally, this product sells 1.8 billion bottles each year—generating enough revenue to sustain a major global corporation, even without all of the other products the company sells. A competitor seeking to capitalize on the global appetite for this type of beverage could invest billions of dollars creating manufacturing and distribution capabilities. Or a single creative entrepreneur could paint a photorealistic replica of the product, and that painting could eventually sell for $57 million and trigger the birth of an entirely new modern art movement. Who is the better businessman, then—the inventor of the soft drink, its many imitators, or the artist who makes an identical copy that can never be consumed?

Andy Warhol did not see the world in a new way; he saw it in the oldest way. He recognized that the image of the Coca-Cola bottle, captured as completely as possible, could expose the contents of the psyche of the 1960s and point to the profound change that had occurred in the American public since the end of World War II. Warhol's paintings of soda bottles and soup cans are iconic representations of the consumerism that had captured the country following the austerity and hardships of the war. Warhol was an innovator who could capture what people were feeling and thinking. His innovation was to forget his own interpretation of the world around him and capture it as it was being created in the mass consumer mind in the mid-20th century. His work was an enormous step away from the abstract expressionist paintings that had captured the art world just a few years earlier.

Warhol's art was a bold, disruptive step into the unknown. It was so separate from and opposed to what art was expected to be that it should have failed. But, like other revolutionary ideas before and since, it touched an entire society, reflecting at a basic level who people were at the time.

Warhol was not alone; all of the modern art movements were led by innovative thinkers who recognized some emerging characteristic of the human condition of their time, and turned it into revolution-ary art.[1] Where does the bravery to take such a disruptive step come from? In Warhol's case, it came from a commercial artist creating advertisements and department store displays in New York City—a man in an adjacent industry looking across the boundary and seeking a bridge to the art world. Warhol wanted to be an artist, but he was poorly equipped to compete with the abstract expressionist works dominating the field, like those of William de Kooning and Jackson Pollock. As a commercial artist, his mastery was in speaking to people in the language of consumer products, with an eye toward sales. He found in that work his pathway into art. He was able to strip away the sales messages from commercial images and present only the visual motifs that were dominating the conversation of the time. Warhol's Pop Art creations were able to say something new about society that could not be said using the expressionists' vocabulary.

Warhol was a disruptive innovator, in every sense of the word. He worked across accepted industrial boundaries. His creations—work that those in power in the art world initially regarded with disdain—ultimately redefined art and shook the art world to its foundations. The gatekeepers could not ignore Warhol's message about society, a message they themselves had missed and did not have the tools to deliver in any case.

We are in a similarly cataclysmic time now, as the rapid advances of technology—from smartphones to cloud computing to 3D print-ing—bring changes in business, in culture, in society similar in scope and impact to those that occurred in the decade after WWII. The impact of these changes is felt every day in the business world, as new companies ascend and old companies decline and fortunes change, sometimes seemingly overnight. This chaotic environment has given rise to a generation of business artists, thinkers developing methods to explain, harness, and benefit from these changes.

Like the pop artists of the 1960s, these innovation artists are creating new schools of thought, not about society and consumerism, but about business models, the relationship between business and society, and innovation. Clayton Christensen has opened the eyes of established companies to the ways in which young startups can gain a foothold in their industries—and eventually displace them. Henry Chesbrough upended the closed-door, proprietary mindset of R&D labs, illuminating the value of teaming with other organizations to leverage multiple domains of expertise. And Eric von Hippel has turned his eye to users, and their potential to act as partners in creating new features and advanced capabilities for next-generation products. When they first appeared, these ideas, and many others like them, ran directly counter to the established thinking about R&D and new product development. Like Warhol, these authors were innovators, creating new models or schools of innovation and new ways to think about the process of creating new products.

Also, like Warhol, these innovators often found themselves developing and maturing in one field before discovering, and acting on, an insight into how their strengths could be applied in a new way. Before he formulated the concept of disruptive innovation, for instance, Christensen worked as a business consultant, government administrator, and corporate CEO. But these experiences did not give him the foundation that he needed to develop his own ideas about business growth. For that, he had to cross into academia, earning a doctorate and a faculty position. Similarly, von Hippel began as an economist and business consultant before shifting his interests to innovation. The traditional path up the organizational ladder seldom includes all the ingredients needed to nurture genuinely new and creative thinking in a field. Unique insights seem to require some time in foreign soil to develop a unique flavor. Innovation originates from a mind able to assimilate foreign ideas, powered by an internal passion to contribute and become successful in a new area.

Viewing the art world through the classic eye of Michelangelo or Rembrandt would suggest that the future lay in improving techniques for realistic representation. But it takes an innovative and disruptive eye to imagine that impressionism, cubism, surrealism, and pop art can all not just exist but make a serious contribution to the evolution of artistic expression. Similarly, looking at a stable and successful business with the classic eye of Frederick Taylor or Alfred Sloan suggests that the only path forward for business lay in optimizing internal efficiency, improving quality, and reducing costs. It takes an innovative eye to see how a product of poorer quality could disrupt an established market leader, or to understand that lead users may be more imaginative in how they actually interact with a product than any internal research department could envision. This is the genius and value of Andy Warhol, Clayton Christensen, and hundreds like them whose perspective is neither easy nor common, but has a powerful impact.

Table 1.—Innovation in art and business

Art Movements		Innovation Concepts	
Date	*Movement*	*Date*	*Concept*
1875	Impressionism	1911	Scientific Management
1905	Expressionism	1970	Strategic Planning
1907	Cubism	1993	Reengineering
1916	Dadaism	1997	Disruptive Innovation
1923	Surrealism	1999	Radical Innovation
1942	Abstract Expressionism	2003	Open Innovation
1954	Pop Art	2005	Blue Ocean Strategy
1960	Minimalism	2005	User Innovation
1965	Superrealism		

Those who seek to make innovative contributions may be advised to guide their career across multiple industries and business functions to accumulate the diversity of information, ideas, and viewpoints that will provide fodder for their passions and engage their talents. Climbing the corporate ladder may require too much adaptation to the existing structure to leave room for, much less foster, a unique and innovative perspective.

For insights into the innovations of Pop Art start on YouTube with "A Guide to Pop Art," https://www.youtube.com/watch?v=LsY4ihZCJL8.

Originally Published in *Research Technology Management*, July-August 2016

CHAPTER 19

MICRO INNOVATION

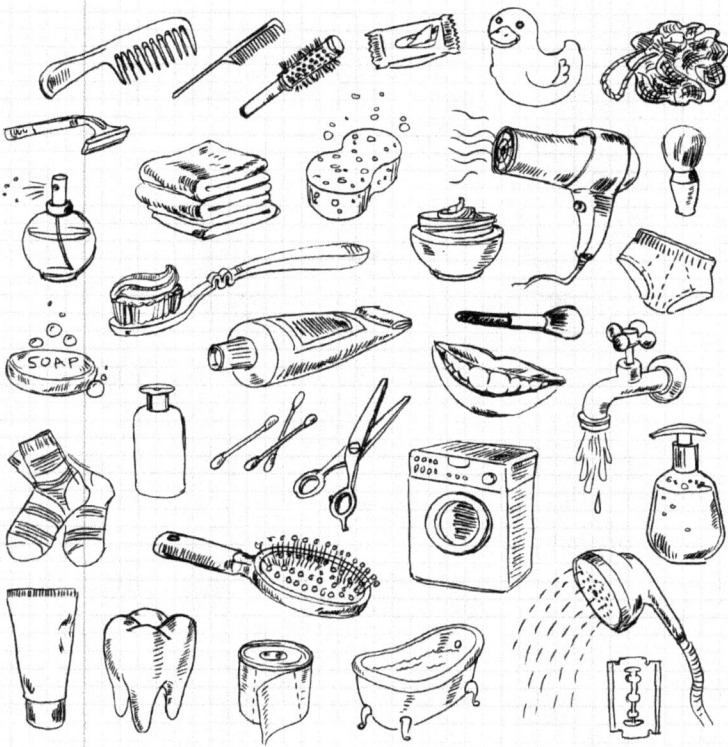

"**W**elcome to the Top Secret R&D lab of Big House Inc. We are working on some of the most confidential and innovative products in the world here. All of you have signed a strict NDA so you cannot reveal anything that you see today. And you will see some mind-blowing stuff, I can tell you that!" Our guide is one of the chief scientists in the company's R&D division.

"Are you making paper out of synthetic materials," I ask?

"No, that would not be paper, Johnny. Follow me and I'll show you the newest product about to be released." The scientist walks the group to a bench scattered with rolls of paper towels. "What do you see here?"

"Paper towels?"

"Right, but these are completely different from anything you have ever experienced. They're going to revolutionize the market and shut down our competitors. Here take a look," he says, handing each of us a roll.

"They feel the same as any paper towel."

"Tear one off and you'll see what we are excited about."

"Oops, I think I made a mistake. This towel is only 6 inches long."

"Exactly! That's where the research comes in! That is the revolution we're going to release on the world. Imagine all of the jobs in your kitchen you need these towels for. How many of those are small jobs like wiping up a dribble of coffee or picking up a dead bug. No one wants to waste an entire 12-inch towel on that job. In fact, some people tear off a towel and then tear it in half, so they can use one half now and save the other half for the next job. No more of that. Now you can use an entire 6-inch towel for those little jobs and get twice as many jobs done with the same roll of towels."

"What? How is that innovation?"

"Think about it, son. That roll of towels in your hand has 200 small sheets instead of 100 big sheets. It can do twice as many jobs.

But the product costs us exactly the same amount to manufacture. We sell it at a 10 percent premium to the standard towels. Our company makes 10 percent in additional profit and the customer gets 100 percent more work from the same roll. This is big innovation. This is like a computer with twice the processing power at only a 10 percent increase in price."

"Really? I don't think that's innovation at all. You just cut the sheets in half and everything else is the same. How is that R&D?"

"I'll tell you how it's R&D, Johnny. It's R&D because no one else has thought of it in the 86 years paper towels have been in existence. None of our competitors, not even Scott Paper, which invented the entire category in 1931, has thought of it. And these little towels are going to take the market by storm. As soon as consumers give them a try, they are going to fall in love. They're going to wonder why someone didn't do this decades ago. They're never going to use a standard-size towel again."

"Ok, so maybe people will like it."

"Let's move on to the next lab. I want you to see the new cap we've created for laundry detergent bottles. I know you're going to get excited about this."

Most of you will find this conversation humorous. How can serious R&D lead to smaller paper towels? Shouldn't R&D be focused on big products like the next cell phone, not just making a household product a little different? That is the common stereotype for R&D and innovation. But it assumes that the thousands of little products we use every day are already perfect for the job they are doing. It assumes that the job they do today is the same as the job they were invented for 10, 50, or 100 years ago. But in reality, consumers, markets, and social structures are always changing. The 12-inch paper towel may have been perfect for the messes of the 1930s, 1950s, and 1970s. But in the 2010s, the messes they are used for are different. Or more likely, the messes have been different for years and the manufacturers could not see a different type of product to address them.

There are thousands of products in our homes and businesses right now that had not changed for decades until a clever and useful new version appeared on the market: Smaller paper towels. A laundry detergent bottle with a cap that drains back into the bottle instead of down the outside. Two-ply toilet paper, dental floss held by a tiny plastic fork, laundry detergent and spot remover in a single-use pod, a razor with three parallel blades (not to mention even newer ones with four and five blades), soda pull tabs attached to the can, portable toast with jelly (Pop Tarts), cooking gelatin with sugar and flavoring added (Jell-O). The list goes on.

By addressing an inconvenience in the way a product is used or enabling a new use, a micro product innovation can revive a solid but tired brand. It can attract customer attention to a line of products that has become almost invisible in the market. It can, in fact, obsolete an entire line of products across multiple companies. The first company to use a liquid laundry detergent bottle with a cap that drains back into the bottle instead of down the side—eliminating the slimy sticky mess on the outside of the bottle—captured the attention and purchasing power of the public almost immediately. The liquid inside had not changed at all, but the bottle addressed an annoying problem everyone had had for decades. Suddenly, no one wanted to buy liquid detergent in the old bottles, tens of thousands of which remained in the pipelines of the inventors' competitors.

Small changes like this can also capture new customers who previously didn't use anything in the entire product category. The clean laundry detergent bottle may have finally converted some users of traditional powder; the convenience of Tide's laundry pods undoubtedly attracted a few more. Dental floss held by a fork may persuade some people to take up flossing, or encourage others to floss more regularly, by making it cleaner and easier to do. The same product in a new form may be a perfect fit for people who were not previously customers.

In other words, one micro innovation can be enough to push an established product to the top of consumers' must-have chart. These innovations may be in the shape of the container (laundry detergent bottle), the size of the serving (half-size paper towels), the convenience of usage (detergent pods), the quantity per package (pocket tissue packs), or a dozen other small features—any of which can address the trigger problems of millions of customers, and shift hundreds of millions of dollars in spending.

R&D and innovation professionals may argue that micro innovation is not different from incremental innovation. But the examples presented here illustrate a definite difference. Incremental innovation typically involves changes to the product itself, to make it better—changing the formulation of a detergent to make it more effective, for instance—or to add features—incorporating fabric softener or stain remover or whitener into the detergent so customers don't have to buy a separate product. The result of an incremental innovation is essentially the same product sold to the same customer base, only modified or enhanced to keep it slightly ahead of or even with competitors. A micro innovation, on the other hand, is a small change, generally to packaging or presentation, that shifts the product so that it solves a different problem—or addresses a problem with the existing product presentation—for existing customers, or addresses a new problem for a new customer.

A micro innovation multiplied by millions of users can equal a significant market share. Procter & Gamble released laundry detergent in single-use pods in 2012. Four years later this new version had captured 12 percent of the $5 billion laundry detergent market in the United States—that's $600 million per year, hardly a trivial innovation. The change might be micro, but the impact can be macro.

This means that R&D investments do not have to be focused on radically new products to realize a significant return. Understanding the usage patterns, unmet needs, frustrations, and preferences of customers can inspire small changes that are valuable to both the customer and the company. Examining features that have been standard for decades can unlock surprisingly large reactions in the marketplace.

In other words, large-scale market impact can be achieved with micro-scale innovation. Small ideas can be truly revolutionary, justifying significant investments of talent, time, and money in changes that appear at first to be trivial.

> Micro, incremental, and peripheral are all adjectives that have been used to explore the power of small, non-core innovations to products. RTM previously published an excellent paper on "peripheral innovation" which readers may want to review for comparison.

References

Bangsil (Esther) Lee and Jina Kang (2016), "The Power of Peripheral Innovation: A Case Study of AmorePacific's Cushion Foundation," *Research-Technology Management* 59(4): 21–29.

Originally Published in *Research Technology Management*, July-August 2017

CHAPTER 20

GOOGLE MEANS *EVERY*

I was teaching an undergraduate math class in 1985 and decided to give the students a fun extra credit assignment. "Find out what a googol is", I told them. This was a time before the Internet and locating such information required a little bit of research in the library. Most chose not to do the research, but a few came back with the correct answer—a googol is the number 10^{100} or a one followed by 100 zeroes.

The original googol term was actually coined by a 9 year-old boy in 1938 and then popularized by his uncle, mathematician Milton Sirotta in his book *Mathematics and the Imagination* in 1940. The number has no real significance in mathematics or the sciences. But it is a useful shorthand notation for the idea of a very, very large number. Math and science teachers sometimes use the googol to give some perspective to the size of the universe or the size of infinity. A googol is larger than the number of atoms in the observable universe, which is estimated at between 10^{79} and 10^{81} atoms, assuming one still believes in the existence of atoms.

The googol has appeared in popular entertainment on-and-off for decades. Charles Schultz used it in a 1963 *Peanuts* comic strip; Steve Martin used it in a 1979 comedy album; it appeared in an episode of the *Teenage Mutant Ninja Turtles*, and was the one million-pound question on the British version of *Who Wants to be a Millionaire*. (Wikipedia, 2009)

But then in 1998, two Stanford graduate students changed the spelling of the word and launched Google, making the obscure mathematics term a household name. The googol was an unknown curiosity for the first 60 years of its life, before getting a makeover and becoming the most widely recognized company brand in the entire world (Jarvis, 2009).

So if googol means a one followed by one hundred zeroes, what does Google mean? Like its namesake, Google means a very, very large number ... of web pages, documents, customers, advertisers, and dollars. Like googol, it falls short of being infinite, but it is difficult to imagine anything else that is quite as big as a googol or a Google.

The company started with a new idea for ranking page searches on the Internet. Instead of giving a high rank to pages on which the search term occurs the most frequently, they ranked a page highly based on how many other outside pages linked to it. This led searchers to the most prominent pages, rather than those engineered with many keywords. The result has been one of the fastest growing companies ever and the expansion of their product line into dozens of information niches. But what do all of the company's niche products have to do with Google's core business, their strategy for the future, and what does this mean to people interested in understanding technology innovation and management?

In short, the mission and strategy of Google is all summed up in the origins of its name—googol. The company is positioning itself to deliver all of the information that goes through everyone's hands, eyes, networks, and hard drives in the future. This kind of universal mission is not new from companies in Silicon Valley. There have been many before Google that have made similar claims. But none of those have had the revenue generation engine to make it happen and none of them have expanded into as many products as Google has.

Google's products are becoming ubiquitous to Internet users. They go far beyond the search service. In fact, the company is really based on advertising today, more than search. Google is using their advertising revenue to create a googol's worth of tools, to reach a googol's worth of users who are performing a googol of different activities.

It is important to look at the tools that Google offers and understand what each of these means in terms of exposure.

Leading Google products and their objectives:

- Search—indexing and delivering **every** web site
- AdWords—used by **every** company to promote **every** product
- Desktop Search—find **every** personal document on your computer
- Images—a doorway to **every** picture on the Internet
- Reader—handling **every** blogged communication thread
- YouTube—streaming **every** video to **every** viewer
- Gmail—delivering **every** email
- Toolbar—presenting Google in **every** browser experience
- Maps—providing directions for **every** trip in **every** city
- Earth—providing a 3D window into **every** part of the globe
- Books—providing access to **every** book
- Picassa—managing **every** digital picture that you take
- Blogger—publishing **every** blog that is written
- Docs—editing and sharing **every** business document
- AppEngine—providing the computer platform for **every** web-based computer program
- Android—initiating **every** cell phone call
- Voice—providing a permanent phone number for **every** telephone customer
- Chrome Browser—becoming the browser for **every** Internet user
- Native Client—running **every** desktop program that you already have
- Chrome O/S—serving as the software foundation for **every** computer

There have been dozens of companies that have attempted to handle everything you do in one niche. But no company has ever delivered every application you will need for everything you do on the Internet. Google is the first to even try.

Google recognizes something that few others seem to understand—everything is a commodity except the information. The computer, its chips, user interfaces, Internet service, and everything else can be copied by a competitor. Given advances in technology, there is always a disruptive innovator around the corner who will be able to build a better product. They are not limited by their investments in old products or their ties to old customers. These disruptive innovators will arise and take the market away from any company that is tied to the past in handling its products and customers. So far, only Microsoft has shown any resilience in holding onto its position. Almost every other company has been seriously challenged or beaten by a smart, disruptive innovator.

Given this environment, the only thing that is not commoditized is the information. Google has set its sights on managing the delivery of all of the information that you see when you are working in the digital world. Every search, every document, every task that you need to do will be handled by Google and one of its ubiquitous and freely provided tools. No other company is operating on this scale. No other company is attempting to have this kind of universal relationship with the world's data and the world's users.

What About Your Company?

Google is moving to the position that every company aspires to—a relationship with every customer in the world. But, how do other companies create a similar information advantage in their industry? If *your* company manufactures industrial tools, how can you create an information rich connection to all possible buyers of those tools? Customers for *your* tools collect data on a number of competitive products and release bids to a number of competitors. How do you insure that they collect the data about *your* tools and invite *your* company to bid?

One way to do this is to become the hub for all information about all products in your market space. Your company web site could collect and distribute all of the information about all of the products that any customer might want. Serious buyers would soon learn that they cannot afford *not* to visit your site. Though they may still collect data from a number of vendor web sites, they would *have to* visit your site because of its reputation for providing all of the data that is available in this area. This is a Google-flavored approach.

Your information hub may provide links to all product specs posted on the Internet. It may host discussion forums open to customers. It may pull in research papers that have been done in universities. It may have comparison tools that allow a visitor to enter the characteristics that they are looking for in a tool and your web site will provide a side-by-side comparison of the specs and costs for all that meet their needs.

Creating such a resource is not really an option open to every company in the industry. Those who already dominate the space are typically large and secure in their position. Their leaders are unlikely to see an advantage in showing all of this competitive data to their customers. Potential customers probably already invite them to every bid and are well educated on the capabilities of their products. Conversely, a company that sells low quality or inferior products is not likely to be interested in highlighting their weaknesses in a side-by-side comparison or publicly accessible customer forum.

Becoming an information hub for the entire industry is really only beneficial to a smaller, up-and-coming company that does not have the attention of all of its potential customers; and whose products can stand-up favorably to comparisons with those of the industry leaders.

Such a web site would potentially rise quickly to the top of the Google search results for relevant terms in your industry. The site would become known around the world, and do so with little investment in advertising and marketing. Once Google finds, ranks, and elevates your information hub, the world will find it. Actions like this could make your company an "every" company for customers in your market. It could insure that you are a "must see" site for any customer who is considering a purchase.

Google means "Every". But it operates in a very universal space. It is not trying to sell to your customers. It is trying to send your customers to the most important web sites based on their queries. If you offer more information than any other company, then you are exactly the kind of site that Google wants to send people too. Their reputation is built on how well they match searchers with the information they are looking for. You can join Google in improving their own reputation by giving Google a genuinely great place to send customers.

References

Battelle, J. (2006). *The Search: How Google and its rivals rewrote the rules of business and transformed our culture.* Portfolio Trade Press.

"Googol" page on Wikipedia.org.

Jarvis, J. (2009). *What Would Google Do?* HarperCollins Publishers.

Vise, D. and Malseed, M. (2006). *The Google Story: Inside the hottest business, media, and technology success of our time.* Delta Press.

Originally Published in *Research Technology Management*, Jan-Feb 2010

CHAPTER 21

WHERE DO THEY FIND THE TIME?

n 1492 while Christopher Columbus was exploring the Atlantic Ocean in search of the Indies, Johannes Trithemius, the Abbot of Sponheim in western Germany, was struggling against the disruptive impacts of the printing press. Johannes Gutenberg had invented movable type around 1439, making it possible to print a book faster than a person could read the book. This famously ushered in a new age of literacy and a transformation of society that included the Protestant Reformation. But, fifty years after its introduction, Abbott Trithemius was facing the concrete effects of the printing press on the monks and scribes whose sole mission in life was to transcribe appropriate materials for the reading public. For fifty years both methods of book production had existed in parallel, but it was becoming increasingly obvious that the life and mission of the scribe was being eliminated one book at a time. Trithemius finally resorted to writing a treatise on the importance of maintaining the life of the scribe in society. He produced *De Laude Scriptorum* ("In Praise of Scribes"), in which he laid out the values and virtues of the scribal tradition. He described four benefits that accrue from writing books versus printing them: (1) the precious time of human life is valuably spent; (2) the scribe's understanding is enlightened as he writes; (3) his heart is kindled with devotion through his writing; and (4) after this life, the scribe is rewarded with a unique prize. (Shirky, 2008a)

Trithemius' treatise focused on the value of transcription to those who are doing the transcribing, not those who need the books. This perspective and concern are common and often repeated when a new technology threatens an old one. The livelihood of individuals is threatened, social foundations are shaken, economic models are broken, and whole classes of people experience unease with the new shape of the world. I described this type of creative destruction in the last column, but this month I want to use these ideas to look at the emergence of the participative web that is occurring all across the Internet, sometimes referred to as "web 2.0" or "the social web."

Until a few years ago, the Internet and the Web were places where established organizations created and published content for the rest of the world to consume, digest, and use in some way. Bulletin boards and private web sites certainly existed, but they were created by a very small minority of technical literati with the unique skills to accomplish what much larger organizations were doing. But then in 1995 Ward Cunningham created the Wiki which was followed by Classmates.com during that same year. These introduced the first social networks and shared web sites. In 1997, Jorn Barger introduced the Web Log (a.k.a. Blog) which allowed people to easily publish a running stream of material in the form of a newspaper, gossip column, or diary. Slowly web users began to realize that they could publish their own Internet content as easily as creating a business document. Ten years later there were millions of personally published web sites. Technorati currently tracks over 112 million blogs, Wikipedia contains over 10 million articles in 250 languages, YouTube contains over 83 million videos, MySpace hosts over 100 million user pages, and Flickr contains over 3 billion photos (Wikipedia, 2008). A modern Abbott Trithemius might ask, "What will this do to the official information providers in society?" and "Where do people come up with the time to do this?"

Content	Current Estimate
Blogs	112,000,000
MySpace Users	100,000,000
Facebook Users	69,000,000
YouTube Videos	83,000,000
Wikipedia Articles	10,000,000
Flickr Photos	3,000,000,000
Ning Social Networks	185,000
World of Warcraft Accounts	11,000,000
Second Life Accounts	13,000,000

Source: Wikipedia encyclopedia, http://en.wikipedia.org/

The answer to the first question is unfolding every day as we witness the loss of consumers of newspapers, magazines, and television broadcasts. One viewer at a time is discovering the interesting, unique, and niche-oriented material on YouTube, Facebook, Flickr, and Wikipedia. The ultimate answer seems clear; the web is supplanting the television as a source of entertainment and information in the same way that the television supplanted radio and newspapers in a previous generation. In the end, all of these will continue to exist, but the balance of social influence and earned revenue will have changed drastically. Paul Saffo has said that, "Silicon Valley is littered with the corpses of companies who mistook a clear view for a short distance." (Saffo, 2007) Though the shift to a new media is clear, the travel time to the future is impossible to determine which makes corporate and venture investment in the new media risky.

The answer to the second question, "where do they get the time", is only obvious to those who are creating the content. The modern "scribal generation" creates products in their professional lives when they are receiving a paycheck to do so, but not as a hobby or as a form of entertainment. This generation has spent the last 50 years comfortably ensconced in front of the television and is convinced that only professionally created content is worth paying attention to. They fail to recognize the local high school football game as entertainment created by amateur hobbyists. This amateur entertainment is not meant for mass consumption, but has a small and devoted following from the classmates and parents of the players. In many ways the social web is like these sporting events. The cost of creation is marginal, it is not initially targeted at a large audience, and the small group that is interested is very devoted. A few of the sites may resonate with the mass populace and become a blockbuster, but most remain small forever. America and the world are becoming more interested in customized entertainment, rather than the mass market entertainment and information that have been in vogue for decades. In a metaphor, we are all beginning to spend more time watching the local high school football team rather than the NFL.

Who is creating all of this content? By one estimate, the people of the United States alone spend 200 billion person-hours watching television each year (Shirky, 2008a). This is a huge investment of mental and physical capital in an activity that produces absolutely nothing beyond a shared consciousness of how Jim and Pam taunted Dwight on the latest episode of a favorite sitcom. The content creators on the web have just redirected a tiny fraction of this consumption time toward the creation of their own content. Mark Wattenberg of IBM and Clay Shirky of New York University estimate that it has taken about 100 million person-hours to create the entire contents of Wikipedia (Shirky, 2008b). If that is in the right ballpark, then a conversion of all television time in the United States could create 2,000 online projects of the magnitude of Wikipedia every year. This would be a huge contribution to shared human knowledge, niche entertainment, and the most complete historical record ever compiled.

Abbott Trithemius was extremely passionate about preserving the scribal tradition and wanted to get his message to as many people as possible, as quickly as possible. His only option was to have the treatise typeset and printed by the very machinery that he was speaking out against. Even his message could not deny the power of the printing press. There is a similar power at work in the spread of participative, social, amateur web tools for creating information content. In spite of all its weaknesses, it is becoming the only way to spread information effectively around the world, and this will soon include commercial business communications. The Gutenberg Bible may be the most famous product of the printing press, but the transformation that it enabled in education and industry in much larger. Social networks, wikis, blogs, and similar tools originated as experimental software and amateur communication outlets. They have already had a significant impact on the news and entertainment industries. They are not going to stop there. These tools will become as central to corporate communications and business operations as e-mail has become over the last decade. Companies that begin to experiment with these technologies will be in a position to leverage them toward competitive advantage just as they did with IT systems in the 1990s. Those who do not will become followers who avoid the early expenses and uncertainty, but who also miss out on the unique advantages that accrue to the leaders. Companies will either align themselves with the tradition of Trithemius or with the innovation of Gutenberg.

References

Shirky, C. (2008). *Here comes everybody: The Power of organizing without organizing.* New York: The Penguin Press.

Wikipedia Online Encyclopedia. Articles on all of the web services referenced in the article. http://en.wikipedia.org/

Saffo, P. (July 16, 2007). "The Future really is now". Computerworld. http://www.computerworld.com/action/article.do?command=viewArticl eBasic&articleId=296816

Shirky, C. (April 26, 2008). "Gin, television, and social surplus". http://www.shirky.com/herecomeseverybody/2008/04/looking-for-the-mouse.html

Originally Published in *Research Technology Management*, Sept-Oct 2008

CHAPTER 22

SUPERCOMPUTING ON YOUR DESKTOP

Ray Kurzweil has garnered a great deal of attention by exploring the future path of computer technologies. His books use past trends and current information to predict what the world will be like ten, twenty, and one hundred years from now. His analysis is so in-depth and his reasoning so persuasive that his ideas are treated as models of the future rather than optimistic speculation. One of his more popular graphs illustrates the date by which a personal computer will have reached a level where it has the capacity to think like a human. He believes that current computers are able to process as much information as a dog's brain and they will reach human levels by 2020 (Kurzweil, 2006).

Quite apart from Kurzweil's predictions, Intel and AMD have delivered computer chips that make major leaps forward in computation every year. Both companies have made it clear that their future processors will offer something different. Instead of doubling or tripling the linear speed of their chips, they will be producing chips that contain multiple processors inside each one (Held et al, 2006). Chips with two or four processing cores are available now, but within a few years they will deliver 32 or 64 processors to a standard consumer desktop. Both companies plan to double the number of processors, also known as "cores", inside of a chip every 18 to 24 months. So for example, if we have 4 core machines in 2008, then we will have 8 in 2009, 16 in 2011, 32 in 2013, 64 in 2015, and 128 in 2017. Within the very short period of nine years we will have gone from a few processing cores to over one hundred. The consumer desktop begins to sound like a small supercomputer. It is a little personal version of a NORAD command center, NSA code breaking machine, Wall Street financial cluster, or Google search farm—right on your desktop.

All of these processing cores present a serious challenge and an industry changing threat to the companies that make the software for these machines. From Microsoft to the one-man programming shop, everyone is going to have to learn to work with this new paradigm of computation. Almost all computer programs are designed to do computation linearly because they have been built to run everything through a single processor. Microsoft Windows® and Office®, the Firefox Bowser, TurboTax®, and computer games are all predominantly a single thread of computation that runs through the computer similar to thread running through a sewing machine. Some programs spin off a few tasks as independent and parallel threads, but there has been limited advantage to this when a computer has only one processor. Eventually everything has to squeeze through that one processor in a linear fashion.

But the personal computers of the future will offer multiple processing cores that do work simultaneously and in parallel. Even though there are 100+ processors in a computer, a traditional linear program will not speed up because it does not know how to spread its work across all of the cores. These multi-core processors require a new generation of software applications that know how to use this power and a new generation of computer programmers who know how to create multi-threaded or parallel computing code. This is a major concern all through the software industry. How do you parallelize the operating system, office productivity software, the web browser, or the next computer game? The companies that master this will be the first to harness the new computing power and deliver "mind blowing" performance to customers. They become the "must have" applications in the next decade.

Multi-processor machines and parallel computing are far from new, they are just not common at the consumer level. They have existed in government, military, and academic products for decades, as well as industrial research labs, but they have never been a staple of consumer-grade equipment nor been part of the average person's experience with a computer. Programs like SETI@Home have allowed anyone to become part of a global, multi-processor computing project that leverages millions of idle computers around the world. But the average consumer has never been able to harness this kind of computing for their own personal needs and entertainment. What does it mean to give this power to every consumer in every home or every employee in every company? How can it be useful?

The recent wave of "web 2.0" applications has given us some clues to what people would do with nearly unlimited access to free computing on a company server and the network bandwidth to access it. Millions of people used these resources to publish their own memoirs, news feeds, diaries, personalized billboards, and encyclopedias. But all of this computation and storage exists in a public space on someone else's computers. It is by nature a public square for public performance, hence the social applications that have emerged with this power. The 128 core desktop machine will be a personal space that you control and can entrust with more private information. Your own personal finances, gambling habits, stock investments, hobbies, family information, and personality can be represented and served.

With every improvement in computer hardware, computer software companies immediately begin to search for ways to exploit this in an application that will attract customers. With 128 processors on a consumer or employee desktop, the possibilities for what can be done with this power are extreme. But it is not clear what the next application with a market the size of MS Office or Internet Explorer will be. During the dot.com boom this was often dubbed the search for the "killer app" (Downes and Mui, 2000). Some killer apps that emerged in the past are shown in Table 7.1. Office applications like word processing, spreadsheets, and presentations have been at the top of this list since they were first invented. In fact spreadsheets like VisiCalc, introduced in 1979, and Lotus 1-2-3, introduced in 1983, are credited with launching the success of the Apple II and the IBM PC as serious business machines rather than niche hobbyist tinker toys. Killer apps are essential because they create a reason for the consumer to buy the machines that can run them. More recently, high-end PC sales have been driven by the demands of visually compelling computer games. Without such games there is little reason to upgrade to a new machine every couple of years. Most applications will run just fine on a five year old machine. Therefore, the killer app has the potential to double, triple, or quadruple hardware sales for the segment of the market that adopts those applications.

Table 7.1. Past and Future Software Killer Apps

Wave	Applications
Office	Lotus 1-2-3, Visicalc, WordStar, Word Perfect, MS Word, PowerPoint
Communication	E-Mail, FTP, Web Browser, News Groups
Entertainment	Digital Video, Digital Audio, MP3, Computer Games
Socialization	Social Networks, Blogging, Wikis, RSS Feeds, PodCasts
Analysis	Weather Modeling, Sports Data Analysis, Personal Rendering, Financial Analysis, Simulation, Digital Buddy

What are the killer apps that will drive the average consumer to purchase a 128 core desktop supercomputers? Two broad categories are generally reliable business areas—entertainment and data analysis. Weather modeling is one example of a universally interesting area for data analysis. The Weather Channel and Weather.com are hugely successful because of the shared interest in the topic across the entire population. Imagine that all of the data that drive these weather sites is available to the average consumer in real-time and that they are offered a program that allows them to easily run their own models of future projections and to explore their own customizations of the data. Imagine that instead of seeing the weather predications for their county, the home user can run predictions for their own city block. What will be the local wind speeds, volume of rain, and accumulated ground water from the latest hurricane in the area? 128 processors on your desktop may allow you to predict this better than your local weather station does today. Consumers just need the tools so they do not have to work with the raw data and write the computer code themselves.

Ever since the 401(k) retirement plan was introduced in 1980, the entire population has become more and more invested in the stock market. These investors watch the market as keenly as they watch the weather. But analysis of stock data is something for the professionals who have access to the data, have created the computer models, and possess the computer hardware necessary to handle these complex models. But given 128 processors, that same data, and the right software, every consumer can run their own analysis of the markets that are as in-depth as those run by brokerage houses today. They can build a picture of their investments and risk levels, and explore potential futures for themselves.

In the entertainment space, fantasy sports competitions attract nearly 30 million people just in the United States. These people's hobby is creating their own professional sports teams and using the performance of real athletes in real games to estimate the outcomes of their own customized teams in fantasy games against each other. These teams and their computations were originally done on paper forms with hand calculators. They are now run on spreadsheets and web servers. The average player uses a spreadsheet, web service, or custom program to study all of the players, build his team, and plan for the future. What if these 30 million people each had access to 128 processors to make their predictions? Would they value a fantasy sports program that allowed them to be 128 times more productive or rigorous than their competitors?

Imagine turning photos and videos into 3D virtual worlds in which the trees in the images are represented realistically in a space that can be navigated. The people in the videos actually walk through the space in the same way that they walk around in the movie. Your son's soccer team becomes a 3D experience that you can watch from any seat in the stadium or down on the field. Military and industrial surveillance videos become a 3D space that can be viewed from all angles. All of this puts the current wave of movie and photo processing software to shame. But it may offer such a compelling experience that everyone has to have the computer hardware and software that can make it happen.

Finally, for the computer gamers, we offer a "digital buddy". Combine the 3D worlds, realistic avatars, artificial intelligence, access to real-time data via the network, voice recognition, voice generation, and lip syn-ching that is available in pieces now to create a computer generated person that is custom built to be your best friend. This is the person who greets you when you come home; asks what you think about last night's football game; tells you that there is a club meeting at the local steakhouse; discusses your views on current events; and listens to everything you want to say, while giving you the kind of feedback you have programmed it for. This sounds impossible now. But each feature required to make this work already exists today and most of them run on the current high-end gaming machines. With 128 processors they could be combined into a single digital buddy that is as real as any person, but customized to your needs. Perhaps this would attract far more customers than even the 10 million players of World of Warcraft that tops the charts right now. The digital buddy might supplant both the computer game and the television as the most popular form of entertainment. Specialized characters could be added to act as business coach, weather man, medical consultant, athletic trainer, or girlfriend. In industry these buddies may replace the training department with customized digital trainers who are knowledgeable on every subject.

Each of these ideas attempts to answer the question—what can I do with the power of a supercomputer on my desktop? It tries to go beyond the current four processor machines and offer applications that have the potential to become more valuable as the number of processors multiplies. For those who can afford the machines, there is a very bright and exciting future ahead as long as the software industry steps-up to create the programs that can leverage this power. Unfortunately, even at consumer PC prices, there are many in society who will not be able to afford these devices and will have no access to the advantages they offer. The "digital divide" in society will become bigger. The "haves" will benefit from more information, better understanding, and more options than the "have nots".

How much electric power will it take to run these machines? We have generally adopted computers and electronic entertainment with little regard to their power consumption. But as electricity becomes more expensive and devices draw more of it, there will be a personal and a social pressure not to expend this power unnecessarily. We may be able to afford the machines and the software, but not the monthly power bill to run them.

From the perspective of the hardware vendors and futurists like Ray Kurzweil, a future with hundreds of personal processors and some really amazing applications running on them lies ahead of us. But other pioneers like Google offer a future driven by online services in which the average consumer needs a less powerful personal computer, not a more powerful one. They suggest that all of the computational, storage, and networking power that you need will reside in "the cloud." If this is the future then computer power will cease to be a consumer product and will become a corporate product offered to consumers as a service. These two opposing visions seem to pit Intel and AMD against Google and Yahoo! in direct competition over the future of computing. The shift to multi-core computing could be as transformative as the first adoption of personal computers in the 1980's. It is clear that the hardware will exist. But it is not clear who will own it or what type of software will be created for the average consumer or professional. The technology creators like Intel, AMD, Google, and Microsoft are wrestling with the very difficult problems associated with creating software for these advanced machines. But the consumer and industrial customers will be faced with the opportunity to do much more complex work on their desktops or online—work that was previously reserved for supercomputers.

References

Kurzweil, R. (2006). *The Singularity is near: When humans transcend biology*. New York: Penguin Books.

Downes, L and Mui, C. (2000). *Unleashing the killer app: Digital strategies for market dominance*. Cambridge, MA: Harvard Business School Press.

Held, J.; Bautista, J.; and Koehl, S., Editors. (2006). From a few cores to many: A Tera-scale computing research overview. Intel white paper at http://download.intel.com/research/platform/terascale/terascale_overview_paper.pdf

Originally Published in *Research Technology Management*, Jan-Feb 2009

CHAPTER 23

COMPUTING IN THE CLOUD

nside a New York City loft in 2006, two entrepreneurs are working on the software to launch a new dot.com business. They hope to offer a service that will allow people to turn a series of photographs into a simple movie with a nice sound track in the background. Though their business offers an online version of something found in most photo viewing applications, they hope that the ability to send the finished product to friends and relatives will make their service a hit with millions of amateur photographers. They have finished the basic functionality of the software, but their next major hurdle is attracting enough venture capital to purchase the server farm and network bandwidth necessary to go live on the Internet. Unfortunately, this will require that they give up significant ownership and control of their company to venture capitalists. But this is how all dot.com companies get started, hoping to make it big like Yahoo!, YouTube, or MySpace.

But, in the 21st century, these entrepreneurs have a new alternative that was not available to start-ups of the past. They can rent the necessary computing, storage, and communication capacity from a large service provider who already has all of these assets connected to the Internet. They can pay only for the volume of these services that they use, they can quickly add or subtract resources from their order, and they never have to take possession of the hardware and all of the technical support headaches associated with it. This will allow them to retain more ownership in their fledgling company, hopefully keeping millions of dollars into their pockets instead of the venture capitalists' pockets. They can launch their company "in the cloud". They can tap into the "cloud computing" services offered by a big vendor like Amazon.com, relying on Amazon's ability to purchase hardware cheaply, maintain it reliably, and staff it competently.

Animoto has become one of the poster children for the cloud computing concept. This producer of online slide-shows was able to launch their company without purchasing millions of dollars worth of computer equipment, hiring an IT department, or selling majority ownership to venture capitalists to fund their company. As a start-up, they really had no idea how big a customer base they could attract, and therefore, did not know how much IT equipment they would need to purchase. But Amazon.com's cloud computing service, known as Amazon Web Services, offered a means to rent this capacity by the compute hour, storage gigabyte, and network gigabit. Amazon also offered to scale up and down as the customer demand increased or decreased. This was perfect for Animoto since they could not predict demand—and since the demand they actually experience varied widely as their service was featured in various press stories. Over one three day period they rocketed from 25,000 registered users to 250,000 as the result of a posting on the popular web site Slash.dot. In a one week period they ramped up their usage of Amazon computers from a couple of dozen machines to nearly 5,000 machines (Figure 8.1).

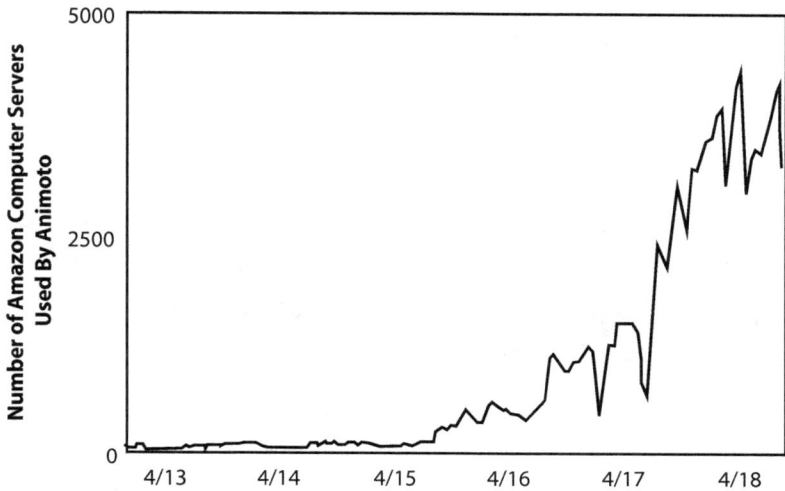

Figure 8.1. Animoto demand for computer servers from Amazon Web Services during April 2008. Source: Information Week, January 23, 2009

WHAT IS IT?

Cloud computing has become a very hot term in the last few years, but a clear description of what it is, what it can do, and why companies might use it is often difficult to find. The concept and a number of predecessor technologies have been around for decades. Gartner, Forrester, and major computer companies offer their own definitions of the term (see definitions box). In essence it is a means of renting computers, storage, and network capacity on an hourly basis from some company that already has these resources in their own data center and can make them available to you and your customers via the Internet.

Nicholas Carr has suggested that computing will follow electricity generation in the pattern of business use (Carr, 2007). A century ago, most companies had to build their own dedicated power generation capabilities. Today, that is performed by an electric utility and companies only purchase the amount of electricity that they need each

Cloud Computing Definitions

Gartner: Cloud computing is a style of computing where massively scalable IT-related capabilities are provided as a service across the Internet to multiple external customers.

Forrester: A pool of abstracted, highly scalable, and managed infrastructure capable of hosting end-customer applications and billed by consumption.

The 451 Group: The cloud is IT as a service, delivered by IT resources that are independent of location.

Wikipedia: A style of computing in which dynamically scalable and often virtualized resources are provided as a service over the Internet.

IBM: A cloud computing platform dynamically provisions, configures, reconfigures, and deprovisions servers as needed. Cloud applications use large data centers and powerful servers that host web applications and web services.

UC Berkeley: The illusion of infinite computing resources available on demand, the elimination of up-front commitments by cloud users, and the ability to pay for use of computing resources on a short-term basis as needed.

day, each hour, and each minute. A century ago, power generation was a barrier to market entry because of the huge up-front cost to build dedicated factory generation capacity. In the late 20th century, the cost of IT was a similar barrier to new companies. This was initially overcome through the use of venture capitalist financing for the equipment. Cloud computing offers to eliminate this barrier by turning computing into a utility that companies can purchase based on specific demand.

Telecommunications went through a similar shift in its structure in the 1990's. Traditionally the telecom providers had installed and sold a hard-wired capacity between fixed destinations. But in the 1990's they sub-divided the capacity, packaged it into Virtual Private

Networks (VPN), and sold volume between destinations that was really composed of many linked segments that had been grouped into a VPN. This was known as the "telecom cloud", the term that evolved into the current cloud computing moniker. (Wikipedia)

Amazon.com has become the leader in providing cloud services to thousands of businesses in the last few years. IT companies like IBM, HP, Sun, Microsoft, and many smaller companies are getting into the business and base their own cloud offerings on the model pioneered by Amazon. Each offer a unique package of services, security, accessibility, reliability, and support that they feel will appeal to the customers they are interested in.

ADVANTAGES

Prior to the popularity of cloud computing, there were a number of related service offerings that attracted only minor attention—i.e. grid computing, utility computing, elastic computing, and software as a service. The basic technologies from each of these has been incorporated into cloud computing and appears to be attracting more interest than these predecessors. This attraction may be a function of the maturity of the technology and the services offered, or it may be driven by a marketing blitz that has occurred as Amazon, Google, Apple, and other big names have gotten behind it.

There are a number of business benefits to this technology.

First, cloud computing is dynamically scalable. Businesses can draw as much computing power as is necessary on an hourly basis. As demand from internal users or external customers grows and shrinks, the necessary computer, storage, and network capacity can be added or subtracted on an hourly basis. Most service providers leave this provisioning up to the customer; though automating this is a valuable advantage that is being pursued by those following in Amazon's wake. Second, the resources can be purchased with operational funds, rather than as a capital expenditure. Many IT departments face a long approval process for capital funding, in addition to the wait for equipment delivery and installation. Cloud computing allows them to bring capacity online within a day and to do so using their operational budgets. Third, the equipment does not reside in the company facility. It does not require upgrades to the electrical system, the allocation of floor space, modifications to the air conditioning, or expanding the IT staff. Computers at Amazon.com consume space, power, and staffing support at Amazon instead of within the customer's company. Fourth, there are competing providers for this service. If the first cloud provider does not deliver acceptable performance, a company can always shift their business to another company offering better service or lower prices.

CONCERNS

Of course, each of these advantages has within it a corresponding disadvantage or concern. First among these is security. Many companies are hesitant to host their internal data on a computer that is external to their own company and that is potentially co-hosted with another company's applications. So far, there has been no client-to-client penetration of software or data hosted in the cloud. That may be due to sufficient security provisions, or it may be because there has been no value in this kind of attack in the past. The second concern is location. Companies may be concerned about the physical location of the data that is being stored in the cloud. The laws of the host country of the equipment apply to the data on the machines. European and Asian companies have expressed concerns about having their data stored on computers in the USA which fall under the jurisdiction of the US PATRIOT Act, allowing the U.S. government to access that data very easily. Third, experienced users of cloud computing services have noticed a big variation in the performance of their applications running in the cloud. When you start an application in the cloud you never know who your neighbors will be on the allocated computer and network. Since many companies are all sharing the resources, it is possible to arrive in a neighborhood that is extremely busy and very noisy, leaving little room for your applications to run and communicate. Fourth, interesting bugs in this large system have yet to be worked out. There have been instances in which entire cloud services have crashed and been unavailable for hours or days. When this happens, your application will be offline until the larger problem is fixed. Fifth, each cloud vendor offers unique services and unique ways to communicate with the computer resources. It is possible for your company to get so deeply embedded into these unique and proprietary services that you cannot move your applications without some major changes to both your software and your data. Sixth, it appears that a cloud provider has an infinite number of computers and storage disks to meet your needs. But, there are a finite number of these resources available and your provider is multiplexing these between the thousands of applications that are starting and stopping every hour. If all customers called for services at the same time, the provider could run out of available resources. This is the cloud computing equivalent of a busy signal on Mother's Day or an insurance claim following a major hurricane.

These concerns, and others that are much more technical in nature, are well known to the major cloud computing providers and the intermediary support companies. All are working on solutions to eliminate them. The real question is, "Can your company increase profits or decrease costs by using cloud computing as it exists now?" and "How will that change as the technology matures?"

WHERE WILL IT GO?

If Nicholas Carr is correct, and computing follows the path of electricity in becoming a utility, then it is just a matter of time before all companies have added cloud computing to the arsenal of technologies that support their internal operations and their external customers. But it is not clear how long this might take. Paul Saffo has said, "Silicon Valley is littered with the corpses of companies who mistook a clear view for a short distance. One of the secrets in my business is that everything changes slower than people imagine. Change only seems fast because people overlook the antecedents. Most ideas take 20 years to become overnight successes" (Saffo, 2007). Even if some technological change "has to happen", it is never clear *when* it will happen. Customers did not jump on utility computing and it is still too early to know how many will really adopt cloud computing, though the numbers appear to be much higher than its predecessors. In support of the adoption of their experience with cloud computing, Amazon offers Figure 8.2 which compares the bandwidth of merchandise buying customers with the bandwidth of thousands of cloud computing applications in their data centers. In mid-2007, they began using more bandwidth for cloud computing than for their traditional retail business.

Cloud computing service providers have made their systems so inexpensive use and easy to access, that there is little reason that companies should not be exploring this option for providing data and services that are not proprietary to their business. There are many forms of internal data exchange, online courseware, and regulatory information that can be hosted and distributed using cloud services. These can give the internal IT department firsthand experience with the ease of use, cost effectiveness, and available functionality of clouds without disrupting or exposing the company's core operations.

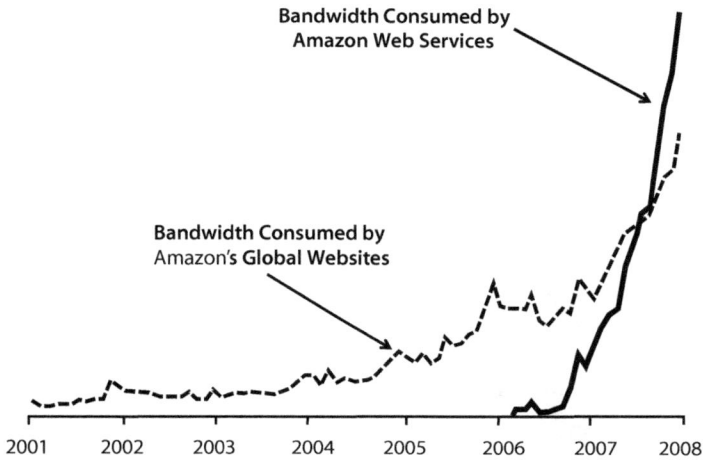

Figure 8.2. Network bandwidth used by Amazon.com retail versus Amazon Web Services. Source: Vogels, 2008

References

Carr, N. (2008). *The Big Switch: Rewiring the world from Edison to Google.*
W.W. Norton & Co. http://www.nicholasgcarr.com/bigswitch/

Paul Saffo. (July 16, 2007). The Future Really is Now. *ComputerWorld.*
http://www.accessmylibrary.com/coms2/summary_0286-33434502_ITM

Vogel, W. (April 15, 2008). Ahead in the Cloud: The Power of
Infrastructure as a Service. Amazon presentation at MySql Conference.
http://mysqlconf.blip.tv/file/981534/

Wikipedia article on "Cloud Computing". http://en.wikipedia.org/wiki/
Cloud_computing

Originally Published in *Research Technology Management*, Sept-Oct 2009

CHAPTER 24

COMPUTING BEYOND THE FIREWALL

The protagonist in Vernor Vinge's 1981 short story "True Names" uses a global communication net to hack into civilian and defense networks, gain control over thousands of computers, and turn them into zombie machines to support his attack on government information stores. "True Names" is about espionage and sabotage on the global information grid, but Vinge's real prescience lies in the story's central assumption—in 1981—that "the network" would become critical to business and government operations and that it would be poorly protected from an intelligent attack *(Vinge, 1981)*.

Thirty years later, we face a regular stream of internet attacks like the most recent in which two virus programs worked together to extract personal, competitive, and financial data from corporate computers. On February 18, 2010 news agencies reported an attack in which the Kneber botnet used the well known ZeuS Trojan Horse to deliver itself to thousands of computers and proceeded to use these machines to collect information and spread itself to other computers. The initial estimate is that this attack yielded 68,000 data records from 2,500 different corporate networks *(Markoff, 2010)*. In spite of the best efforts of corporate IT managers, information security consultants, firewall engineers, and encryption specialists; criminals and state-actors continually find ways to take the information they want. This is a natural consequence of putting computers on a global network. Machines on a network are extensions of that network. The technologies that tie together the insides of a stand-alone computer are not that different from those that extend it into the global network. So it should be no surprise that remote pieces of a networked computer can interfere with the operations of our personal or corporate machines.

The dividing line between private computer, private network, and public internet is fading fast. The shift toward a globally distributed and openly accessible computer grid has been underway for years. It has grown so subtly that we have taken the early manifestations for granted. We have reached a point where most of the work we do is dependent upon and distributed between multiple users and organizations. We are already operating as part of a large network that is intentionally open to group members from various organizations as determined by the work we are sharing. We access data from the internet and corporate networks and share all of our own work via these networks. We have all become nodes in a very large Open IT infrastructure that is not clearly defined or protected.

The world of Open IT has begun. It is being executed every day by thousands or millions of employees who are just trying to get their work done as efficiently as possible. Hundreds of services to help them do this are quickly and easily available. No corporate IT policy or firewall can stop this proliferation of data without also crippling the operations of the company.

This may have begun in the 1990s, when computers and networks became powerful enough to allow employees to telecommute. Companies created virtual teams whose members resided around the country or around the globe, coming together only through the exchange of electronic data. Enabling such distributed work required trusting all of the computers and networks with some portion of the corporate data necessary for the work to get done. IT managers may have installed encryption software and virtual private networks to assure everyone that the data was secure, but the extension of the corporate network through the open internet and into employees' privately owned machines meant that the system was open to intrusion from other users of that same network. The boundary between corporate, public, and private information devices had been taken down and the resulting open system offered an attractive target to those who saw the value of stealing information or disrupting its free exchange. Mobile communication and mobile computation opens this door even further.

How is this migration to Open IT happening and why is it necessary? Increasingly, no company can provide all of the tools that can be found on the internet and no company can survive without access to those tools. Ask yourself a few very practical questions about where you get the information services necessary to do your job.

- Where do you get the maps and directions that you use to drive to a meeting across town? MapQuest?
- How do you share documents with partners in your own and other companies? Google Docs?
- Where do you store customer contact information? Salesforce.com?
- How do you access the most current world news? NYT Online?

In each of these cases, you are using a service created for easy access to anyone that wants it. The service is generally not contracted specifically and securely by the corporation, but rather is selected for its convenience and usefulness by each employee. What is the result? Is your corporate customer list stored at Salesforce.com or LinkedIn? Do your confidential collaborative documents reside in Google Docs? Is corporate data being analyzed on servers owned by Amazon or another cloud-based provider? In most cases, the decision to do any of these is in the hands of each individual employee, not an all-powerful IT czar.

Each employee usually has the power to use these services from their personal computer or their corporate computer.

Open IT in which data and computation is distributed all over the Internet is the logical end-state of these kinds of operations. This trend will only continue to grow as our need for services and our reliance on networked operations increases. If this is inevitable, we must determine how it can be carried out securely and with some level of trust. In his 2008 novel *Halting State*, Charles Stross's characters wrestle with a global network in which all data from banking to government has been subdivided, encrypted, and distributed to be hosted and computed on machines everywhere on the internet. Security is based on the theory that it is impossible for any attacker to identify the location of any single important piece of data, pull together thousands of small pieces, and break the unique encryption on each of them. In this story, there is no central data server that belongs to a single company. The entire internet is the data center for every organization and it is protected by fracturing valuable data into small pieces, distributing them anonymously, and wrapping them in trusted encryption. All business and personal operations are protected and executed in this ultimate Open IT environment, and everyone in the world trusts this system based on the encryption, distribution, and anonymity that the scheme provides *(Stross, 2008)*.

Halting State's concept of distributed, but secure, corporate operations across the entire internet is the logical end-state when corporations and employees need many more services than can be provided internally. It appears to predict the universal use of the "Web 2.0" services that have been appearing in recent years. As quickly as IT departments expand their services, the needs of companies expand even faster. Employees looking for more efficient ways to do their jobs find answers on the internet, spreading corporate data and computation across that open network.

Every time a successful internet attack is reported in the news, there is a renewed effort in governments and corporations to lock down networks and ban certain kinds of applications. This reaction is contrary to the direction that business operations are moving. Government, corporate, and personal activities rely increasingly on the use of the network to store, access, and share information. We can't prohibit or build fences around one of the most powerful inventions in human history. Instead, we must find a way to make internet-based operations secure. The work we are doing on the network is too important to be done without protections. All communications and data exchange need the kind of encryption and user authentication that Stross describes in his novel. The alternative is to accept an infinite series of successful attacks against our poorly protected systems, operations, and data.

Just as the global information grid described by "True Names" in 1981 has become a reality, the Open IT environment described by *Halting State* in 2008 is already on its way. Our modern systems have to evolve away from the vulnerable design of Vinge's 1981 world toward something more like the more secure world of Stross's novel.

References

Vinge, V. (1981). "True Names", a short story appearing in *True Names and the opening of the cyberspace frontier*, edited by James Frenkel. Tor Books, 2001.

Markoff, J. (February 18, 2010). "Malicious Software Infects Corporate Computers". New York Times Online. http://www.nytimes.com/2010/02/19/technology/19cyber.html?hp

Stross, C. (2008). *Halting State*. Ace Books.

Originally Published in *Research Technology Management*, May-June 2010

CHAPTER 25

R&D IN THE FINANCIAL CRISIS

R&D at NASA, NSF, NIH, and DOE has fared very well in the Obama administration's stimulus plan. But what role does R&D really play during a financial crisis? Where does it fall on the list of economic priorities for the country when there is a long list of organizations that need financial support?

R&D funding certainly creates jobs for scientists in the same way that infrastructure funding creates jobs for construction workers or defense funding creates jobs for engineers and manufacturing workers. But which is more immediately stimulative - a dollar spent in a research lab or a dollar spent on the highway? Both investments provide jobs today and finished products tomorrow. Research funding might lead to a new synthetic material or energy source in 10 years. A new highway or bridge might lower transportation costs and increase economic activity between regions that lie along its path.

The effects of R&D spending on the national economy were the topic of a 2007 study by the U.S. Bureau of Economic Activity and the NSF. This study showed a direct growth in GDP for industrial sectors that rely heavily on R&D (Okubo, 2006). Until recently, economists have measured R&D in the same way that they have measured infrastructure, the cost to create a physical facility (Mandel, 2008). But economists are trying to change this. They have realized that the historic link between innovation, R&D, and new jobs seems to have stopped (Mandel, 2008). In the 2009 edition of the annual NSF R&D survey, 40,000 businesses have been asked to describe how they use internal and government funds for R&D, where it is spend and how much is directed toward services versus manufactured products (NSF, 2009).

There is a need for an economic model that describes the national stimulative effect of money spent on R&D. Certainly the proponents for R&D and infrastructure can both provide case studies of investments that bore no fruit, while both can also show cases where small amounts of money led to huge economic returns. But is there any dependable pattern to the effect of spending in these areas?

Lawrence Summers, the new Director of the National Economic Council, has said that, "[a] question that should be much more of a preoccupation for all of us is—what are the animating technologies that are going to drive our economy forward? There was a tremendous wave of innovation in connection with the Second World War: the jet airplane, electric technologies, and more. Their diffusion fueled a period of rapid productivity growth, which improved standards of living and made almost everything else work well for the generation after the war." (McCormack, 2008)

Summers is focusing on the ROI for investments in new technology. R&D prior to, during, and after WWII provided huge advances in radar, microwave, radio, television, nuclear power, and transportation. The research generation that followed brought us personal computers, cellular communication, the Internet, web-based businesses, and logistic and operational efficiencies. But the WWII era investments were not completely unique in time and effect. That initial momentum of invention and innovation has carried through to this day. The 1940's were a significant launching point that changed forever the nation's and the world's views of what science and technology can add to human existence.

Our own publication has included many articles on the positive, negative, and null effect that R&D investment has on the financial future of individual companies within an industry. But we do not appear to have looked at the impact that R&D has on the national economy and compared it to the impact of other activities and investments. This kind of insight and a model of it are essential tools in engaging national leaders to apply public money to R&D or creating R&D incentives for corporations.

The economic crisis and the election of a new president have led a number of writers to offer prescriptions for R&D.

David Goldston has suggested that, "The Obama administration's promised economic stimulus package offers another opportunity to align policy goals with research priorities. Truly inventive transportation research has never received more than crumbs. We need more R&D on information networks and intelligent highways that direct drivers to the fastest routes, better-planned communities that reduce the need to drive in the first place, and more flexible and appealing mass transit systems." (Goldston)

His focus on transportation suggests improvements in efficiency and quality of life. Investments in infrastructure, which may include new technology, can reduce the amount of wasted time and energy that the nation spends commuting between home and work. These changes could improve individual and organizational productivity, getting more work out of the same resources. This kind of productivity was one of the major interests of former Federal Reserve Chairman, Alan Greenspan. He could see the impact that computer technologies were having by allowing each person to generate more value and revenue for a company and for the country. Turning "drive time" into "work time" could have a similar effect, though the scales may be different. Greenspan believed that computerization was improving productivity by 3% per year, effectively adding 1.2 hours of productive labor to each employee every week for a year (Greenspan, 2007). Reducing drive time may be able to generate this kind of improvement as well, but without a compounding effect. It might eliminate one hour of daily travel from those in the new highway and community system, providing a 12.5% fixed productivity opportunity for the fraction of the population that are significant commuters with access to the new system. This 12.5% boost is a one-time, but constant level of improved productivity. Its effect on the national economy is diluted by the ratio of commuters who can use it to those who do not commute or cannot access the system. There may be some logic in assuming that people who have such a significant commute hold jobs with high pay and high contributions that motivate people to accept the commute in the first place. So this improvement may accrue to exactly the people who can make the biggest contribution by adding an hour to their day.

David Duncan observed that President Obama was very supportive of R&D funding and eager to find areas in which research can make a big difference in society. "During the campaign, Obama talked about launching an initiative to create renewable sources of energy akin to the Apollo space program in the 1960s that put men on the moon. Among other things, he said he wants to double the budgets of the National Institutes of Health, the National Cancer Institute, and other federal research and development agencies in the next 10 years." (Duncan, 2008)

Improvements in medicine are another form of productivity enhancer. In a society with a larger portion of senior members, medical advances that allow them to live a more healthy, active, and productive life accrue to the benefit of all of society. Instead of suffering the infirmities of age, a big portion of society may remain healthy enough to avoid heavy use of government supported medical programs. Some of them can choose to continue working and making an economic contribution at an age when their parents were unable to make that choice. These kinds of economic contributions are not of the accumulative type described by Greenspan, but are more like an annuity that makes payouts every year, and is reducing in value rather than growing through compounding.

Finally, Julian Siddel has reflected on the new administration's intention to create a Federal CTO position. "I think that's very important in an era when most of the nation's major policy challenges revolve around science and technology. The president can't make fully informed policy decisions without taking the science into consideration, so having that voice in cabinet discussions is very important." (Siddle, 2009)

In the face of financial crisis this seems to be an administration that understands the importance of science and technology in society and the historical impact that successful R&D has had on society. But, in the absence of an economic model that can compare the beneficial effects of a dollar of research versus a dollar of infrastructure, the question of which is better for the country falls away unanswered.

From 1975 to 1980 the Industrial Research Institute operated its own Research Corporation in conjunction with member companies, the NSF, DOE, and DOC. Cooperatively the group studied the effects of corporate R&D and specific types of technology. After 20 studies and two million dollars, activities of the corporation were suspended. This demonstrates a precedent of working with government agencies which could be reemployed around areas of the national economy.

References

Okubo, S. et al. (2006). Bureau of Economic Analysis/National Science Foundation R&D Satellite Account: Preliminary Estimates. Online at: http://www.bea.gov/newsreleases/general/rd/2006/pdf/rdreport06.pdf

Mandel, M. (Sept 11, 2008). Can America invent its way back? *Business Week Magazine.* Online at http://www.businessweek.com/magazine/content/08_38/b4100052741280.htm

National Science Foundation. (2009). Business R&D and innovation survey. Online at http://www.nsf.gov/statistics/srvyindustry/about/brdis/

McCormack, R. (Dec 3, 2008). Research and development will not be included in President-elect Obama's massive economic stimulus package. *Manufacturing and Technology News,* 15(21). Online at http://www.manufacturingnews.com/news/08/1203/summers.html

Goldston, D. (no date). Science we can believe in: How President Obama can recharge US research. *Wired Magazine,* 17(1). Online at http://www.wired.com/culture/culturereviews/magazine/17-01/st_essay

Greenspan, A. (2007). *The Age of turbulence: Adventures in a new world.* Penguin Publishing.

Duncan, D. (Nov 5, 2008). Natural Selection. *Portfolio Magazine.* Online at http://www.portfolio.com/

Siddle, J. (Jan 20, 2009). Scientists optimistic over Obama. *BBC News.* Online at http://news.bbc.co.uk/1/hi/sci/tech/7792171.stm

Originally Published in *Research Technology Management,* May-June 2009

PART V

INNOVATION LEADERSHIP

CHAPTER 26

EINSTEIN AND PICASSO IN R&D

On April 18, 1955, Albert Einstein, the world's most famous scientist, passed away. The mystique around his intelligence was so great that the doctor performing the autopsy gave special attention to the brain. In fact, Dr. Thomas Harvey removed Einstein's brain, weighed it, photographed it, dissected it, and preserved the pieces. These pieces have traveled the world to be studied by numerous scientists who are eager to find a connection between Einstein's unique intelligence and the physical structure of his brain.

Albert Einstein

But there was another genius contemporary to Einstein who received very little scientific scrutiny after his death. Pablo Picasso, born just two years after Einstein, changed the art world with Cubism and its influence on modern art. But when Picasso died, no one performed an autopsy on his brain. It was not dissected or studied to determine what made him such an artistic genius. Why not? He was equally influential and brilliant in his own field.

Picasso

Picasso's genius wasn't scientific, so he didn't capture the attention of the scientific world in the same way that Einstein did. But Picasso's artistic genius may be just as important to the effective operation of an R&D department as Einstein's scientific genius.

MULTIPLE INTELLIGENCES

Genius comes in many forms. Howard Gardner (1983) suggests that there are multiple, unique forms of intelligence; he identified a total of eight "intelligences":

- Logical-Mathematical—scientific and technical talent;
- Verbal-Linguistic—the ability to use words and language effectively;
- Interpersonal—the ability to interact effectively with people and teams;
- Intrapersonal—self-reflective and self–understanding tendencies and talents;
- Visual-Spatial—imaginative and artistic talent;
- Bodily-Kinesthetic—physical talent and dexterity;
- Musical—the ability to create music; and
- Naturalistic—an ability to manage and relate to the natural world.

We all have some mix of these intelligences, while most of us balance several. The exceptional brain may express itself through math and science—as Einstein's did—but it may also excel in language, relationships, spiritual understanding, art, even physical abilities.

We tend to recognize most readily those geniuses who share our own intelligences. In the R&D department, we respect the logical-mathematical intelligence—and we may miss the value of other forms of intelligence. Is there a place for other forms of intelligence as well in the R&D lab? How would an R&D department built by Picasso differ from one built by Einstein?

EINSTEIN'S R&D DEPARTMENT

Einstein would almost certainly staff his R&D department with the smartest scientists and mathematicians. Colleagues like John von Neumann, Stanislaw Ulam, and Werner von Braun, all of whom showed their expertise in creating the atomic bomb and exploring new theories of physics, would be at the top of his recruitment list. Such a brilliant group would seem to be the natural choices for an R&D department; surely, there would be no theoretical or technical problem that they could not solve.

Einstein might also recognize the need for interpersonal skills in the managers who would oversee and organize the scientists. His experience at Princeton's Institute for Advanced Studies showed him that scientists could be very contentious, unwilling or unable to compromise without effective intermediation. With this in mind, Einstein the R&D architect might turn to someone like Robert Oppenheimer, who was a master at dealing with these kinds of personnel problems on the Manhattan Project.

With these two categories of skills accounted for, Einstein may well close the door to his department and set off to create new products for the likes of General Electric, General Motors, ALCOA, IBM, or AT&T. But would such a department be successful from a business perspective? Would all of this logical intelligence, guided by talented management, be able to create products that were both functionally valuable and aesthetically attractive to customers?

PICASSO'S R&D DEPARTMENT

Pablo Picasso's R&D department, on the other hand, would likely be composed primarily of visual-spatial geniuses like himself. He would probably recruit other artists, perhaps the masters who started him down the path toward Cubism. Edouard Manet, Claude Monet, Camille Pissarro, and Edgar Degas could be the leading thinkers and creators in Picasso's department. Such a unique, creative group should be able to out-design any competitor. They may not create advances in technology, but they would significantly improve the visual and functional style of everything they touched.

Picasso was a sculptor as well as a painter; as a result, he may also recognize the role of bodily-kinesthetic intelligence in product testing and usability. Perhaps he would seek out some of these geniuses for a department that would be able to suggest functional modifications to the design team. Such an R&D team would be the pride of companies like Apple, Google, and Hyundai.

EINSTEIN + PICASSO

Most likely, neither of these geniuses alone would create the optimal R&D department. In either case, the resulting structure would be lopsided, heavy in the area of the founder's personal genius. While both Einstein and Picasso represent the extreme success of one unique form of intelligence, their stories in isolation could lead to a neglect of the benefits of other intelligences.

The truly successful R&D department will use all eight intelligences (Figure 11.1). Executives and market-facing staff members who promote new products and services and attract attention to them need verbal-linguistic genius. Interpersonal geniuses can handle internal personnel and organizational issues, while intrapersonal geniuses identify the psychological needs of customers. The visual-spatial geniuses would create the look and shape of a product, making it sexy and attractive. Bodily-kinesthetic geniuses test product functionality and ergonomic fit. And the logical-mathematic geniuses create new technology that is the basis of completely new products.

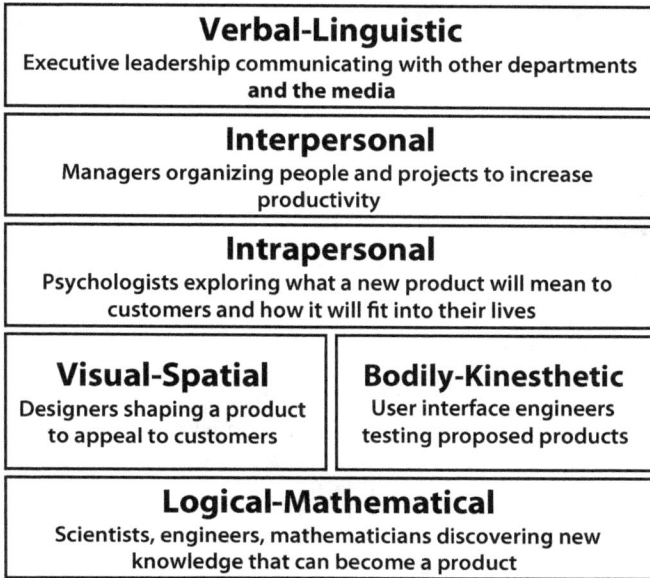

Figure 11.1. Different Types of Intelligence in R&D

Something like an Einstein vs. Picasso contest seems to be playing out in a number of industries, but most noticeably in the Apple vs. Microsoft battle. Each company creates the operating system for a major proportion of the world's computers. Both have chosen to pursue the creation of hardware, Apple in digital music (iPod®) and cell phones (iPhone®) and Microsoft in computer gaming (Xbox360®). But while Microsoft has made huge investments in R&D, Apple has captured a much bigger market with an R&D budget just under a quarter the size of Microsoft's (Peers 2011). In this case, it seems that Picasso's team has created a product with a much larger market impact, something that is ubiquitous that offers significant visual-spatial advantages over competitors, not to mention a distinct design signature. Meanwhile, the Einstein team has focused on overpowering competitors with a product that is impressive to hard-core technologists, but not terribly relevant to the largest portion of society. Is this because Einstein cannot see that mass consumers need a ubiquitous product that is perhaps less feature-heavy than the Xbox360®, while Picasso is not attracted to the raw power of computer technology when there are such obvious opportunities in design?

Is it easier to be an Einstein or a Picasso? Does Picasso's R&D come from a brain that is somehow more unique than Einstein's brain? When we create R&D teams, are we putting too much emphasis on that which is most easily measured and too little on less easily recognized forms of genius, which may have greater impact?

Does the culture inside of a company allow an R&D team to choose to be Einsteinian, or Picassoesque, or a combination of the two? Or does corporate culture dictate which pattern must be used?

References

Gardner, H. 1983. *Frames of Mind: The Theory of Multiple Intelligences*. New York: Basic Books.

Peers, M. 2011. RIM: Less research = more motion. *Wall Street Journal*, March 30, C16.

Originally Published in *Research Technology Management*, Sept-Oct 2011

CHAPTER 27

UNDERSTANDING AND ACQUIRING TECHNOLOGY ASSETS

INTRODUCTION

Technology has become an integral part of nearly every business and social endeavor. However, in spite of this, each profession has different definitions for what technology is. A universally shared definition has not emerged—which indicates that the transformation of these professions by technology is still occurring faster than it can be codified.

A physical scientist might describe technology as the set of equipment and apparatus that are used for scientific experiments. A social scientist would make a more vague reference to the underlying change agent that is advancing society. An IT professional sees technology as the computer hardware and software that is used to automate internal business operations. A manufacturing plant manager might suggest that technology refers to all of the assets that enable and enhance production operations. An economist sees technology as an enabling force in society that can make significant improvements to productivity on a global scale. The diversity of these perspectives is an indication of the pervasiveness of technology, and the challenges associated with understanding how it impacts business and social activities.

Burgelman, Christensen, and Wheelwright (2004) define technology as,

> "the theoretical and practical knowledge, skills, and artifacts that can be used to develop products and services, as well as their production and delivery systems. Technologies can be embodied in people, materials, cognitive and physical processes, plant, equipment and tools. Key elements of technology may be implicit, existing only in an embedded form (like trade secrets based on know how) and may have a large tacit component." (p. 2)

Christensen (2003) defines technology as,

> "the process that any company uses to convert inputs of labor, materials, capital, energy, and information into outputs of greater value. For the purposes of predictably creating growth, treating 'high tech' as different from 'low tech' is not the right way to categorize the world. Every company has technology, and each is subject to these fundamental forces." (p. 39)

Porter (1985) insists that,

> "technological change is one of the principal drivers of competition. It plays a major role in industry structural change, as well as in creating new industries. It is also a great equalizer, eroding the competitive advantage of even well-entrenched firms and propelling others to the forefront. Many of today's great firms grew out of technological changes that they were able to exploit. Of all the things that can change the rules of competition, technological change is among the most prominent." (p. 164)

UNDERSTANDING TECHNOLOGY ASSETS

Prahalad and Hamel emphasize the importance of integrating technology assets in order to develop the core competencies of the organization, "core competencies are the collective learning in the organization, especially how to coordinate diverse production skills and integrate multiple streams of technologies" (Prahalad and Hamel, 1990). But they do not detail what these streams of technologies are.

In their 1994 book, *Competing for the Future*, these same authors state that, "a core competence is a tapestry, woven from the threads of distinct skills and technologies. ... Many companies have had difficulty blending the multiple streams of science or technology that comprise their heritage into new, higher-order competencies" (Hamel and Prahalad, 1994, p.214). Again they identify the importance of technologies, but assume that the manager will be able to identify all of the streams of technology that are important to his business.

Sharif (1995 and 1999) suggests that the streams of technology referred to by Prahalad and Hamel fall into four major categories and that mastering these technological assets is essential for competitively positioning a company (Figure 12.1). These comprise the "THIO Framework":

- Technoware—object-embodied physical facilities
- Humanware—person-embodied human talents
- Inforware—record-embodied codified knowledge
- Orgaware—organization-embodied operational schemes

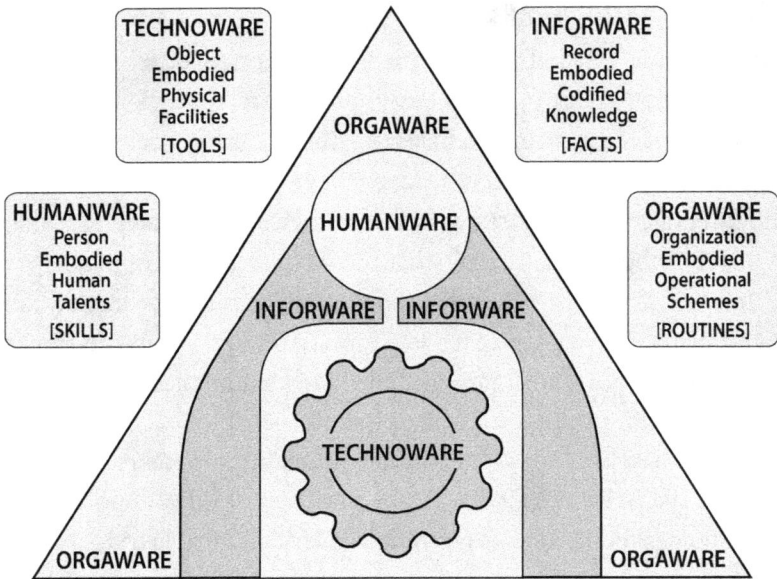

Figure 12.1. There are four technological components that play an essential role in creating and establishing a competitive position for a company. Source: Sharif, 1995

Technoware refers to equipment, laboratories, and other assets that a company can acquire or create to assist in creating a product or offering a service. Humanware refers to the capabilities of the people in the organization and their ability to apply those capabilities in a productive manner. Inforware is the knowledge that is encoded in documents and processes and that are accessible to the organization. Finally, orgaware describes the capabilities of the organization that are derived from its structure and the processes that determine how it operates.

Christensen and Overdorf (March-April 2000) wrestle with this same issue of defining the valuable assets of an organization when they discuss its resources, processes, and values. They emphasize that the capabilities of new companies are often concentrated in their people (i.e. humanware) because operational processes and organizational values have not had time to form yet. The resources of a start-up company may also include technoware in the form of unique equipment or patent protection on a new technology. Christensen and Overdorf's "processes" are an expression of Sharif's orgaware in that they refer to the valuable capabilities of the organization as unique from both individual people and specific equipment. Their "values" capture the organization's analysis of the industry and market (i.e. inforware) to determine what they will specialize in. Subramaniam and Youndt (2005) also recognize the importance of humanware and emphasize that it is one of the essential ingredients for enabling radical innovation in an organization. They go to state that the social relationships between people are an equally important ingredient for innovation—a.k.a. orgaware or social capital. Finally, their research indicates that patents and historical knowledge/information within the organization create organizational capital (i.e. inforware) that is an essential ingredient for enabling incremental innovation of existing products and services. The technology start-ups in Silicon Valley are classic examples of the importance of humanware at the beginning of a venture. The algorithms that established Google as the leading search engine in the world were created and implemented by its two founders, Larry Page and Sergey Brin. In the beginning, their expertise was the most important ingredient in making the company successful. However, over time, that skill and knowledge is not sufficient to grow and operate the business. The company must add organizational capabilities, supporting technologies, and protection of their proprietary information.

In their report on the need for innovation in America, the Council on Competitiveness (December 2004) emphasized the differences between small start-up and large established companies, specifically that small companies rely on the depth of expertise of individuals (humanware) while larger companies rely on the capabilities of the organization (orgaware) and often lack the ability to access unique individual expertise.

Sharif also accepts that there are financial and natural resources available which are not necessarily related to technology. The importance of natural and financial resources was also emphasized by Daniel Bell in describing the evolution of society from its agricultural roots, through its 19th century manufacturing foundation, to the more recent post-industrial or information economy (Bell, 1973). The pattern of this evolution is shown in Figure 12.2.

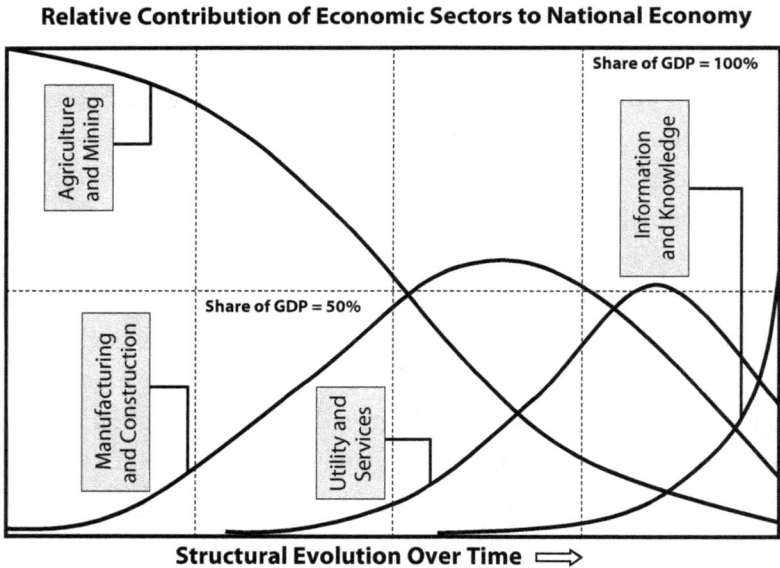

Relative Contribution of Economic Sectors to National Economy

Share of GDP = 100%

Agriculture and Mining

Information and Knowledge

Share of GDP = 50%

Manufacturing and Construction

Utility and Services

Structural Evolution Over Time ⟹

Figure 12.2. The contribution of specific resources and industries to the social economy has evolved over time.

Leonard-Barton (1992) suggests that there are 4 dimensions (or assets) that make up the knowledge-set that enables technological innovation (Figure 12.3). These are:

- Skills and Knowledge Base—knowledge and skill embedded in employees (i.e. Humanware)
- Technical systems—knowledge embedded in technical systems (i.e. Technoware)
- Managerial systems—formal and informal ways of creating knowledge (i.e. Orgaware)
- Values and Norms—traditions from the founders (i.e. Inforware)

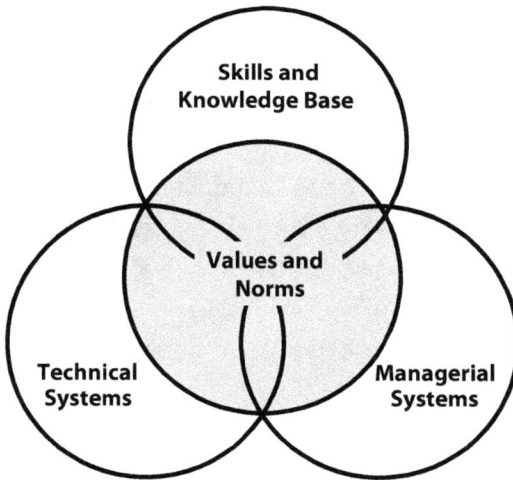

Figure 12.3. Leonard-Barton presents four dimensions of knowledge that contribute to organizational capabilities. Source: Leonard-Barton, 1992

Each of the above models of technological assets presents them as interlocking or interdependent. The authors emphasize that an organization needs all of them if it is to be successful. Different types of industries and competitors may need them in different proportions, but few if any industries totally omit any one category. The balance between these assets also varies over time as a company or industry matures or is transformed by changes in its social or technical environments.

ACQUIRING TECHNOLOGY ASSETS

"For the past 25 years, we have optimized our organizations for efficiency and quality. Over the next quarter century, we must optimize our entire society for innovation."
— Council on Competitiveness, 2004

As industry and business have evolved, so have the means of acquiring technology to improve the effectiveness and productivity of human economic endeavors. Chandler (2001) identified the essential role of an "integrated learning base" in supporting the long-term success of a company's innovation programs. He studied the emergence of the consumer electronics and computer industries to determine why some companies were extremely successful in the short-term, but completely lost their position over time. The learning base to which he refers is synonymous with the acquisition of new technologies and internal human expertise in applying these technologies. The ability to apply these within a company was one significant differentiator between those companies that survived as producers of electronics and those that have failed because they had no internal, integrated competitive advantage. Chandler offers RCA as a classic example of this pattern of growth and decline. RCA held patents for a number of important electronic components and manufacturing methods. However, instead of developing internal capabilities to turn these into products, the company chose to make its money by licensing these patents to other manufacturers. As a result, RCA never developed their own internal expertise with the technology and when advances came they came from other companies, leaving RCA holding the rights to outdated technology. It was primarily the Japanese companies that had licensed RCA IP who developed the next generation of technologies and methods and used them to pull the entire electronics industry away from American companies and into Japanese companies (Chandler, 2001).

In a knowledge economy, the ability to acquire, organize, and apply new knowledge is an essential ingredient in effective innovation. Christensen and Raynor (2003) state that, "corporate IT systems and the CIOs who administer them figure among the most important contributors to failure in innovation" (p.89). They emphasize that ready access to useful and customized information is a prerequisite for growing and changing the organization. Inforware is not just a controlled body of knowledge that enters with a new employee or is created through internal activities. Rather it is a global environment that extends far beyond the boundaries of the organization. Ingesting and managing as much inforware and technoware as possible is an important part of stimulating the innovation process. This occurs by sending employees to trade shows, industry trade publications, experimenting with competitors' products, looking for inspiration in other industries, and a number of similar approaches to information acquisition.

Technoware is often seen as being the most sophisticated because it has the power to do more work. Its efficiency in reducing materials and energy, its ability to contain self-guidance and control, and its ease of use and reduced impact on the environment place it in the limelight of executive attention. But without humanware, this technoware is nonfunctional and useless. Though technoware often contains encoded humanware, it still requires human control and application, which requires human knowledge and skill. Additionally, Orgaware is the structure that is able to bring together the right technology and human skills with a market opportunity. Shifting IBM from a company primarily focused on computer hardware to one centered on services required pulling together all four of the resources we have identified. They had to master the technology of the new open systems environment, which required applying their humanware expertise, technoware capabilities, and inforware licenses. They also had to recreate the organization so that it could operate effectively around services.

Start-up	Expansion	Consolidation	Leadership
Humanware	**Technoware**	**Inforware**	**Orgaware**
Human resources are dominant. Values are beginning to form.	Acquisition of technology resources to expand the business and improve productivity.	Understanding of competitive environment and selection of identity based on values.	Creation of organizational structure and processes.
			Competency focuses on the creation of effective organizational structures and the alignment of business processes.
Competitive advantage stems from the unique skills of individuals and small groups.	Technology assets and equipment add to the competencies of the people and expand the market reach of the company.	Mastery of information about the industry, customers, suppliers, and government lead to specialization.	
			Organization applies its significant resources in accordance with the business processes and organizational structures that encode its operations.
Organization has minimal established capabilities to support competencies	Technology assets create an initial foundation for corporate capabilities beyond human capital.	Organization establishes processes to govern its resources and to allow them to become independent of uniquely talented individuals.	

Competencies (vertical axis, left) — *Competencies & Capabilities Ratio* — *Capabilities* (vertical axis, right)

Leonard-Barton (1992)			
Skills & Knowledge	Technical Systems	Values and Norms	Managerial Systems
Christensen & Overdorf (2000)			
Resources (Human)	Resources (Technology)	Values	Processes
Subramaniam & Youndt (2005)			
Human Capital	Organizational Capital	Organizational Capital	Social Capital

Figure 12.4. Different technology assets make different types of contributions to the growth and competitiveness of a company as it moves through its lifecycle.

Figure 12.4 brings together the concepts presented by multiple authors to illustrate how each of them describes a similar phenomenon. These assets are aligned with the business growth phase, showing which asset is dominant during each phase of the company's lifecycle.

During the start-up phase of a company, the humanware assets are the most important ("skills and knowledge" in Leonard-Barton and "resources" in Christensen & Overdorf). The unique skills of individuals are usually the foundations of the company's ability to compete in an industry. As the company expands, it uses its financial capital to purchase technoware that will allow it to extend the productivity of is humanware and reach a larger market ("technical systems" in Leonard-Barton and "resources" in Christensen & Overdorf). As it runs into stiff competition and its expansion caries it into areas that it cannot excel at, the organization realizes the need to consolidate. It must apply inforware about the market, customers, suppliers, and competitors to determine what its unique place should be. This selection leads to a definition of its values and norms, the definition of what it will pursue and what its measures of success will be. If the company survives it can potentially enter the market leadership phase of its lifecycle. In this phase, its orgaware is most important. The creation of an organizational structure that can operate the business independent of the individual humanware and technoware assets that were the foundations of the company is essential ("managerial systems" in Leonard-Barton and "processes" in Christensen & Overdorf).

Throughout this evolution, as a category of technological assets moves from a dominant position to a supporting position, it also moves from being a competitive competency of the organization to being an operational capability. These assets are always important, but they become woven into the fabric of the company, to use Prahalad's analogy, and are part of the stable foundation of capabilities rather than the front-end transformative force of the organization.

APPLICATION TO GLOBAL COMPETITION

The four categories of technology assets are essential resources both in defending current market positions and in usurping those positions from competitors. Burgelman asserts that, "From a competitive strategy point of view, technology can be used defensively to sustain achieved advantage in product differentiation or cost, or offensively as an instrument to create new advantage in established lines of business or to develop new products and markets." (Burgelman, 2004, p.143) Having achieved an advantage, technology assets are one essential ingredient in defending that position. Operational efficiencies are necessary, but these can be copied. The earlier quote from the U.S. Council on Competitiveness emphasized the need for innovation to remain competitive, and their report focused on the application of new technology and investment in R&D as a key part of innovativeness.

Christensen and Raynor (2003) attempt to identify actions that senior executives must take to lead this innovation. These actions align very well with the THIO framework that is at the center of this paper. First, executives should stand astride of the interface between sustaining and disruptive innovation for their organization. They should examine the threats of new technology (study and apply technoware) and the need to maintain the capabilities of the current organization (foster humanware). Second, they should champion new processes for generating disruptive growth (advancing orgaware). Third, they should sense when circumstances are changing and teach others to recognize these signs as well (monitoring inforware and mentoring humanware).

von Hippel (2002 and 2005) and Chesbrough (2003) both point to an additional dimension of this model for managing innovation, one that extends beyond the boundaries of the company. von Hippel points out that there are "leading-edge users" of every product. These people and organizations press the product to its limit and often end up inventing modifications that are beyond what is delivered in the original product. As the broader consumer base for these products evolves, it will also discover a need for the modifications pioneered by leading-edge users. Therefore, a company needs to tap into these leading-edge users, create partnerships with them, and bring their modifications into the product research and design process. Chesbrough's ideas concerning "open innovation" talk to the need to leverage the capabilities of multiple organizations to create new products. He has observed that no company possesses the expertise necessary to innovate in all of the domains that apply to its products. Therefore, partnerships are necessary to maintain a lead over more insular competitors. These ideas extend the management and optimization of the THIO framework beyond the boundaries of a single company.

Early Starter Advantage

Technological advancements form S-curves in which early applications provide small, incremental improvements, but these soon lead to significant or exponential improvements. During this exponential phase, it is tempting to believe that the technology will continue to improve business operations, productivity, efficiency, and cost savings at this rate. But as the potential within each improvement is realized, its contributions taper off significantly to an incremental tail. When an industry is in a stagnant phase, it will remain in the incremental improvement tail for a significant period. Luckily, complex organizations have a number of opportunities to apply new technology assets and to jump onto the early phases of a new S-curve (Moore, 2005).

This makes it very important for a company or a country to adopt and apply new technologies early enough that the explosive financial benefits are still available to pay for start-up costs, which may be significant. If a company or country waits too long to apply a new technology, then it may find itself in a position where the profits available cannot overcome the start-up costs.

Once a company is established in an industry, it can benefit from multiple waves of technological improvement. Moore (2005) points to the importance of applying innovative technologies throughout the lifecycle of the company. During some phases it is possible to innovate in the technoware components of the product. During others it is possible to innovate in the orgaware/production processes. At other times it may be necessary to innovate in the humanware domain.

Innovation may emerge in many different parts of the organization, but it is unlikely that transformative changes will be continuous in any one area. Instead, disruptions in one area will be followed by stability and standardization to make those changes into a repeatable part of the organization's operations. While this standardization is occurring, disruptive innovation may emerge in one of the other domains that drive productivity and competitiveness (Christensen, 1999). Microsoft has experienced the early starter advantage and has had to wrestle with the disruptions that have occurred within and around its operating system business. Having successfully captured the desktop operating system market, Microsoft still had no control over the evolution of the definition of the operating system from the customer's perspective. Companies like Qualcomm and Netscape extracted the email client and the web browser from university labs and introduced them to Windows® customs. Microsoft missed the opportunity to introduce these tools themselves and had to catch-up to the idea that they should reside side-by-side with every copy of the operating system. More recently, this early starter is facing the same challenges from search engines, media management programs, media editing suites, blogging tools, and a growing list of contenders who hope to create the next ubiquitous tool for the Windows environment.

Late Starter Advantage

Not all industries require a major investment to enter—i.e. they do not have a significant barrier to competitive entry (Porter, 1985). When this is the case, it is possible for a late starter in the field to have an advantage over early starters. Early starters typically pay a premium price for equipment that is just being created to take advantage of technological advancements. Early starters also take the largest risks in predicting market demand and experimenting with new production processes. Since technoware changes so rapidly, it is possible that the early starter will spend significant money and time pursuing failed products and markets. This may make it possible, even advantageous, for another company to start later, but hit the right market with the right product the first time out. Under these conditions, the late starter may outperform the early starter and capture a dominant position in the market (Markides & Geroski, 2005). Apple's iPod® is a fantastic example of this approach. They entered the MP3 player market five years after many of the early starters. They had the advantage of understanding the approaches of dozens of existing competitors and most of the necessary technology had already been created. Apple brought two new ingredients to the MP3 device—a massive internal hard drive that could store thousands of songs and a superior user interface that appealed to a larger portion of the consumer market. These two advantages allowed them to consolidate a fractured market, capture 70% of the business, and redefine what an MP3 player should be.

CONCLUSION

In this paper we have attempted to clearly identify the types of technology assets that a company must acquire and apply in order to be successful in the marketplace. Numerous authors have talked about the importance of managing technologies and "weaving streams of technology" without explicitly defining these technologies. Referring back to Christensen's definition of technology as "the process that any company uses to convert inputs of labor, materials, capital, energy, and information into outputs of greater value" (Christensen, 2003), we suggest that managers must consider much more than just traditional R&D and the acquisition of new equipment that represent "hard technology". Rather, a manager must leverage the power of humanware, technoware, inforware, and orgaware as described in this paper. Further, we believe that each of these plays a dominant role during a different phase of a company's lifecycle. As an asset moves from a dominant position to a supporting position, it moves from a differentiating competency, to an operational capability. A company cannot survive without creating a strong foundation of capabilities. But capabilities can often be duplicated by competitors, so it is difficult for them to continue to provide a competitive advantage. Therefore, a company must continue to innovate with new technology assets.

References

Bell, D. 1973. The Coming of post-industrial society: A Venture in social forecasting. New York: Basic Books.

Burgelman, Robert A., Christensen, Clayton M. and Wheelwright, Steven C. 2004. Strategic Management of Technology and Innovation, (4th Ed.), Chicago, IL: Irwin Publishers.

Chandler, A. 2001. Inventing the electronic century: The Epic story of the consumer electronics and computer industries. New York: Free Press.

Christensen, C. and Overdorf, M. March-April 2000. "Meeting the challenge of disruptive change". Harvard Business Review.

Christensen, C. 1997. The Innovator's dilemma: When new technologies cause great firms to fail. Boston: Harvard Business School Press.

Christensen, C. 1999. Innovation and the General Manager. Boston, MA: Irwin McGraw-Hill.

Christensen, C. and Raynor, M. 2003. The Innovators solution: Creating and sustaining successful growth. Boston, MA: Harvard Business School Press.

Chesbrough, H. 2003. Open Innovation: The New imperative for creating and profiting from technology. Harvard Business School Press.

Council on Competitiveness. December 2004. "Innovate America: National Innovation Initiative Report—thriving in a world of challenge and change". U.S. Council on Competitiveness. Accessed August 2, 2006 at http://www.compete.org/

George, M., Works, J., and Watson-Hemphill, K. 2005. Fast Innovation: Achieving superior differentiation, speed to market, and increased profitability. New York: McGraw Hill.

Hamel, G. and Prahalad, C.K. 1994. Competing for the future. Boston, MA: Harvard Business School Press.

Leonard-Barton, D. Summer 1992. "Core capabilities and core rigidities: A Paradox in new product development". Strategic Management Journal, 13, pp. 111-126.

Markides, C. and Geroski, P. 2005. Fast second: How smart companies bypass radical innovation to enter and dominate new markets. San Francisco, CA: Jossey-Bass.

Moore, G. 2005. Dealing with Darwin: How great companies innovate at every phase of their evolution. New York: Portfolio Books.

Porter, M. 1985. Competitive Advantage: Creating and sustaining superior performance. New York: The Free Press.

Prahalad, C. and Hamel, G. May-June, 1990. "The Core competence of the corporation". Harvard Business Review, pp.79-91.

Sharif, N. 1995. "The Evolution of technology management studies: Technoeconomics to technometrics". Technology management: Strategies and applications for practitioners, 2(3), pp. 113-148.

Sharif, N. 1999. "Strategic role of technological self-reliance in development management". Technological forecasting and social change, 44(1), pp. 219-238.

Stalk, G., Evans, P., and Shulman, L. March-April 1992. "Competing on capabilities: The New rules of corporate strategy." Harvard Business Review, pp. 57-69.

Subramaniam, M. and Youndt, M. June 2005. "The Influence of intellectual capital on the types of innovative capabilities". Academy of Management Journal, 48(3), 450-463.

von Hippel, E. 2002. "Innovation by user communities: Learning from open-source software". in Innovation: Driving product, process, and market change. ed. E. Roberts. San Francisco, CA: Jossey-Bass.

von Hippel, E. 2005. Democratizing innovation. Boston, MA: MIT Press.

Originally Publish in *Technovation*, April 2007

CHAPTER 28

THE FIELD-GRADE CTO

I f a chief information officer (CIO) can manage the internal use of information technology throughout an organization and a chief financial officer (CFO) can oversee the finances of an entire company, then it seems logical that a chief technology officer (CTO) should be able to direct the use of all non-IT technologies across a company's product-development and manufacturing processes, doesn't it?

But technology is not finance, or even IT. Technology is both more diverse and more specialized than finance and IT, and it may be more difficult to manage with the same top-down hierarchy used in those domains. Within any large corporation, there are literally hundreds of unique technologies to be evaluated, adapted, and incorporated into products or production processes. An aerospace company may have interests in metals, composites, radar systems, and avionics. At an oil company, the central technologies may be in remote sensing, seismology, and oceanography. While they are related areas, they are also widely divergent. It is difficult, if not impossible, for a single CTO to get his arms around all of the technologies that may be important to a complex organization and provide meaningful guidance about which ones to pursue and how.

Lewis and Lawrence (1990) counseled the CTO to get out of the research lab and contribute to the business strategy: "The CTO's key tasks are not those of lab director writ large but, rather, of a technical businessperson deeply involved in shaping and implementing overall corporate strategy." Perhaps, I would like to suggest, the CTO should also get out of the C-suite and into the detailed workings of the business units. In a world of diverse technologies, there is a need for more senior technologists looking into fewer technologies each. What's needed, in short, is a field-grade CTO.

FIELD-GRADE OFFICERS

The military has a long history of embedding functional experts into their field units. These field-grade officers fall between the senior ranks of generals and the lower company officers who have direct command of the troops. The field officer does not directly manage and direct combat troops; rather, he or she focuses on a specialty area such as logistics, intelligence, or communications, along with all of the details involved in that field. He or she brings a unique expertise to the field units, adding specialized knowledge where and when it is needed. Similarly, "field-grade CTOs" could be distributed across business units, providing specialized expertise in the few technologies most important to each unit.

In fact, this structure has already been adopted in many companies, where technology leaders are focused on the use of technology within a specific field unit of the company. Having served as a CTO in a software company, a government acquisition office, and a nonprofit hospital system, I have observed that the function of the CTO has become much more of a field operation, rather than a single C-suite position (Smith 2007). Though there may be a single CTO at the top, most companies also employ a number of business-unit level CTOs engaged in the operations of just one specialized area. Further, this field-grade CTO may or may not have official reporting or accountability relationship to the C-suite CTO.

David Pratt, for example, has served as the chief technology and engineering officer, chief scientist, and fellow for the modeling, simulation, and training business unit of SAIC, a company of 46,000 people headquartered in Northern Virginia. He reports to the business unit's senior vice president and weighs in on all strategic issues involving that unit's products and services. He has a voice in decisions to make acquisitions, pursue new contracts, and expand into new markets. But he does not serve as the CTO for the entire company, nor is he expected to be a master of every technology that this global company uses. His relationship with the C-suite CTO is more akin to those within a consulting company. The C-suite CTO does not control the daily activities of Pratt and his peers in other SAIC business units, or even evaluate their annual performance. Rather, he expects to be able to call on the specialized expertise of the field-grade CTOs when new problems arise across the company. If a new business opportunity requires expertise in virtual reality combined with global communications, the company can call upon the expertise of Pratt and other experts to deliver a solution more rapidly than competitors. Once the immediate problem is solved, the field-grade CTOs return to their business units. What hierarchy that does exist among the CTOs is not meant to control their daily activities, but rather exists to provide a competitive advantage to SAIC in new business ventures.

PORTER'S TECHNOLOGY VALUE CHAIN

In his classic book, *Competitive Advantage* (1983), Michael Porter Michael Porter stated that, "technology is embodied in every value activity in the firm, and technological change can affect competition through its impact on virtually any activity (p.166). Most managers and executives remember this as the book that introduced the value chain, but they may have missed this discussion of technology across that value chain.

In his value chain model, Porter divided the company's various operations into primary activities—inbound logistics, operations, outbound logistics, marketing and sales, and service—and support activities—infrastructure, human resources management, technology development, and procurement. In addition to describing how each of the activities, whether primary or support, create value, Porter offered a concise description of the role of technology within each area. For instance, operations technologies include the materials that are used in production, the machinery that is used, the means of handling and transporting material down the line, packaging for intra-facility handling, quality assurance methods, facilities design, and information technology on the production line.

Seemingly anticipating the clamor for a single C-suite CTO that would arise through the 1990s, Porter showed how technologies and their applications within different activities of the business are unique and each unique application makes significant contributions to the value chain. A field-grade CTO in an operations area, for instance, will focus on core manufacturing processes and materials. How is the product created and how can it be done faster, cheaper, and better?

And support activities are not a technology backwater. In fact, the IT department has totally transformed Porter's infrastructure, procurement, and human resources management activities since the book's publication. Leaders have redesigned entire companies around new IT infrastructures, introducing changes on a par with the addition of electric power and telephones in an earlier generation. IT is now so important that the CIO position has become almost universal in the executive ranks of companies around the world.

Not surprisingly, many of these CIOs are supported by a deputy with the title of CTO. A CTO-of-IT, like Herb Keller at Florida Hospital, usually handles the relationships with specific IT product vendors and the actual installation of the tools. Executive guidance from the CIO establishes the funding and strategic plan, which the

CTO-of-IT then implements as part of the infrastructure support activity. He must also work with the operating units of the hospital, such as the emergency department, surgery, in-patient care, and the research group, to meet their unique IT needs. When the CTO of the research group wants to establish global telecommunications connections to conduct research into telesurgery, he and the CTO-of-IT develop a CTO-to-CTO collaboration between two field-grade officers to find effective, affordable, and regulation-compliant tools that can do the job. Though these CTOs have different reporting structures within the organization, they share the responsibility for solving an important problem within the hospital system.

CONCLUSION

Any title with the CxO form is automatically expected to be part of the executive suite, with one person holding the title. But the CTO title may not fit that mold. Given the variety of technologies—and contexts in which technologies are implemented—it may be time to rethink the notion of a C-suite CTO. Porter's Value Chain model suggested the need for these technology experts as early as the 1980s. Industry practice has met this need through the creation of field-grade CTOs.

References

Lewis, W. W., and Lawrence, H. L. 1990. A new mission for corporate technology. *Sloan Management Review* 31(4), p.57-67.

Porter, M. 1985. *Competitive Advantage: Creating and Sustaining Superior Performance*. New York: The Free Press.

Smith, R. 2007. What CTOs do. *Research-Technology Management* 50(4), p.18-22.

Originally Published in *Research Technology Management*, May-June 2011

CHAPTER 29

ALIGNING COMPETENCIES, CAPABILITIES AND RESOURCES

Innovation in technological competencies, organizational capabilities, and the application of resources is a necessary prerequisite to maximize a company's ability to penetrate the market with new products and services. In this paper we extend the work of Prahalad and Hamel (1990) and other authors to demonstrate the importance of aligning innovations in these three core areas. This alignment is illustrated with the analogy of an axe penetrating and splitting wood. The paper illustrates the difference between innovations that are aligned and supportive of a common goal, as compared to organizations in which these three components are independent and not supportive of each other.

"**C**ore Competencies are the collective learning in the organization, especially how to coordinate diverse production skills and integrate multiple streams of technologies" (Prahalad and Hamel, 1990). The introduction of core competencies had a major impact on management practice and thinking. Multiple authors adopted, adapted, and extended the ideas of core competencies. One of the most prevalent adaptations was to change "competency" to "capability" and apply a more general definition to the term. Stalk, et al (1992) stated that, "whereas core competence emphasizes technological and production expertise at specific points along the value chain, capabilities are more broadly based, encompassing the entire value chain." They go on to propose that core capability is "a set of business processes strategically understood" and that it represents "technological and production expertise at specific points along the value chain." Leonard-Barton (1992) turned core competency into core capability in this way, "core capability is an interrelated, interdependent knowledge system." Even Hamel and Prahalad sometimes use the terms interchangeably in their later writings (1994).

In this paper I propose that there is an important distinction to be made between competency and capability. Providing different definitions of these two terms is valuable in aligning two different sets of practices within a company. This alignment is essential to the effective penetration of the market with new and existing products. I propose that capabilities refer to a broad set of practices in which a company has proficiency. But that these practices are rooted in production and daily operations. A capability is the organizational ability to execute activities repetitively, efficiently, and predictably.

Contrasting this, a competency refers to a company's ability to improve its performance continuously. A competency is the source of differentiation for the company allowing it to create and offer unique products, services, and solutions to customers. A competency is the organizational ability to improve continuously.

Further, established companies possess many more capabilities than they do competencies. They have developed the ability to execute repetitively in a number of areas. But they have relatively few competencies, or areas in which they are able to improve their performance continuously. Contrasting this, new companies have relatively more competencies and fewer capabilities. Their entire business strategy is based on a few things that they can do differently than established industry leaders, but they possess very few capabilities to deliver products and services repetitively and efficiently.

MARKET PENETRATION

In order to penetrate the market, a company must be able to align innovations in both its capabilities and its competencies for the effective satisfaction of customer needs. Established markets are filled with products that meet the needs of a specific set of customers. New entrants into the market must provide either a better product or a different product in order to displace those that already exist. Porter emphasized two sustainable strategies of entering and remaining in an industry. A company must be able to offer the same products at a lower cost, or they must be able to offer differentiated products that cannot easily be duplicated by competitors (1985). Christensen extended this perspective by demonstrating the power of technological advancements to enable a low-cost strategy to be transformed into a differentiated product (1997). Christensen's disruptive innovation brings out the power of technology to create major competitors from companies that previously would have been permanently relegated to the role of a niche player.

Given the opportunities presented by low cost, differentiated products, and technology disruption, a company must structure itself to deliver these advantages consistently, repetitively, and efficiently to customers. Without a complementary strategy across the company, a new product or service cannot be pressed forward to create a permanent and growing position in the market.

In addition to competencies and capabilities, a company must align its resources to feed the production and management systems that deliver the volume and quality of products needed. Essential resources include personnel, technology, information, finances, and natural resources.

INNOVATION ALIGNMENT: THE AXE ANALOGY

Companies have competencies, capabilities, and resources. All of these must be aligned to be effective in penetrating the market. Without such alignment, a product or service might have sufficient financial resources, but insufficient production capability. It may have world-class manufacturing capabilities, but poor R&D and innovation competencies to create new products. Applying resources, capabilities, and competencies individually or without alignment is not an effective strategy for market penetration.

This idea is illustrated with the analogy of an axe splitting wood (Figure 14.1). The wood represents the market that is to be penetrated. It is dense with existing products and services. There are also interlocking relationships among products because a customer uses many of these together. In order to enter this market, a new product must provide a better solution and it must be able to break existing bonds.

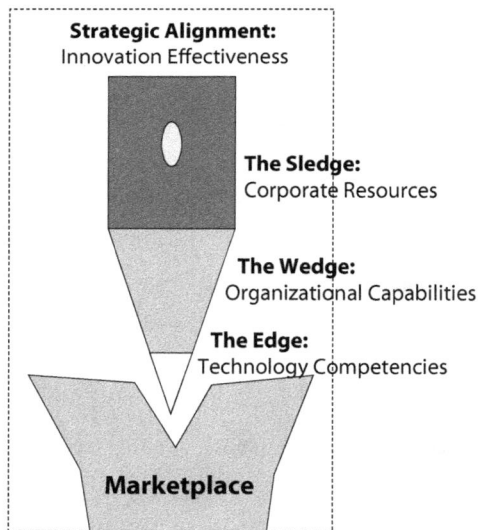

Strategic Alignment:
Innovation Effectiveness

The Sledge:
Corporate Resources

The Wedge:
Organizational Capabilities

The Edge:
Technology Competencies

Marketplace

Figure 14.1. An axe penetrating wood is an apt analogy of the power of strategically aligning competencies, capabilities, and resources to maximize market penetration.

The sharp edge of the axe blade represents the core competencies of the company to create a better product. The edge is honed through research and development, the application of new materials, the creation of new state-of-the-art production capabilities, or the application of products from an adjacent industry. This sharp edge penetrates the market and separates established product relationships.

Separating established relationships is not sufficient for taking market share. Following the edge there must be an organizational wedge that is designed to push aside competing products and replace them with the new competitor's products. The wedge represents the capabilities of the company to continuously deliver the products and services. This includes manufacturing, logistics, marketing, partnerships, labor relations, and a host of other capabilities to follow-up on the disruptive entry of the edge of the axe into the market.

Finally, the sledge represents the resources of the company to continue to feed competencies and capabilities. The resource sledge includes the people, factories, logistics systems, natural resources, and finances necessary to push the edge and wedge deeper into the market, opening a wider space for the new competitor's products and services.

Using this analogy, we can also demonstrate the limitations associated with applying any one of these individually. Resources alone deliver a blunt object against established products and relationships. It is like chopping wood with a sledge hammer, it may dent the surface and disrupt some small part of the market, but it will not penetrate (Figure 14.2a). Large oil, gas and gold producers are heavy users of resources that could attempt to enter a new market by applying only the brute force of their resources.

Competencies alone can penetrate the surface and break some relationships, but without capabilities, this will make only a small cut in the wood. There is no wedge behind the edge to open a significant space for the new products (Figure 14.2b). Many R&D-focused

start-up companies are based entirely on competencies. They have excellent skills in a narrow area, but lack the ability to apply them effectively, such as through effective marketing, distribution, customer service, or information processing.

Capabilities alone do not possess the edge to break into the market or the resource sledge to deliver significant force behind the blow (Figure 14.2c). A large, low-cost manufacturing company typically has significant capabilities, but without either unique competencies or abundant resources. As noted, all three must be aligned to effectively penetrate the market.

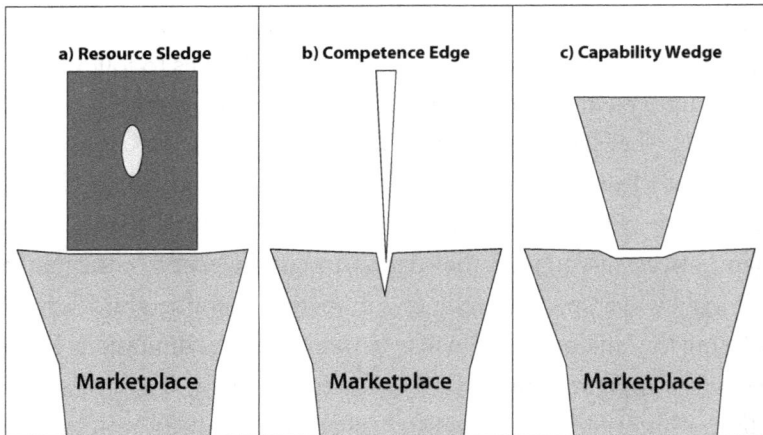

| a) Resource Sledge | b) Competence Edge | c) Capability Wedge |
| Marketplace | Marketplace | Marketplace |

Figure 14.2. Individual innovations in resources, competencies, or capabilities are significantly less effective at market penetration.

COMPETENCIES: THE EDGE

In their 1990 HBR paper, Prahalad and Hamel state that, "core competencies are the collective learning in the organization, especially how to coordinate diverse production skills and integrate multiple streams of technologies." They go on to emphasize that core competence should: (1) provide potential access to a wide variety of markets, (2) make a significant contribution to the perceived customer benefits of the end product, and (3) be difficult for competitors to imitate. (Prahalad and Hamel, 1990)

Further, in their 1994 book, *Competing for the Future*, these same authors provide a much more distinct definition that is more useful to us in differentiating competencies from capabilities. They state that, "a core competence is a tapestry, woven from the threads of distinct skills and technologies. ... Many companies have had difficulty blending the multiple streams of science or technology that comprise their heritage into new, higher-order competencies" (Hamel and Prahalad, 1994, p.214).

These new higher-order competencies refer to a company's ability to improve continuously. Investments in R&D are one traditional method of continuous improvement. To remain relevant and valuable, competencies must be renewed and changed. They must be able to make "significant contributions to perceived customer benefits". If competencies are not renewed, then the customer will move away from the solutions offered yesterday toward better solutions offered by new competitors.

Sharif (1995 and 1999) emphasizes that a company's competencies must include an ability to seek out solutions, to ask questions, and to experiment with new ideas. It cannot limit itself to better efficiency with existing products and processes (a capability). Competence is "solution seeking" and requires the synthesis of ideas from many domains and time periods. It looks beyond what is practical, feasible, profitable, and immediately approachable. This aligns well with Christensen's description of the emergence of a disruptive product from roots that at first appear to be inferior to current solutions. The key is that the new roots have much greater future potential than the old roots. Seeing this, appreciating it, and pursuing it requires the freedom to look beyond current capabilities.

Organizational learning is an important ingredient in maintaining a competency. The organization must be able to absorb and integrate multiple streams of knowledge (Prahalad, 1998). They must be able to share this knowledge across the organization such that it can move from where it is discovered, created, or appreciated to where it can be effectively applied. In many companies, strong organizational boundaries have the effect of fracturing core competencies because they separate complimentary knowledge, prevent communication, and disincentivize collaboration (Leonard-Barton, 1992).

Inside the organization, there must be entrepreneurs who are able to pursue new and innovative paths. These people must "learn to forget" (Prahalad, 1998) about established practices and seek out new solutions (Sharif, 1995). Over time, these groups must even learn to forget about established competences. When a competence no longer meets customers' needs or cannot be extended further, it does not provide competitive advantage. Continuing to adhere to these exhausted competences is a "competence trap" (Levitt and March, 1988) or a "core rigidity" (Leonard-Barton, 1992)

CAPABILITIES: THE WEDGE

"Whereas core competence emphasizes technological and production expertise at specific points along the value chain, capabilities are more broadly based, encompassing the entire value chain." (Stalk et al, 1992, p. 66) Capabilities are "a set of business processes strategically understood ... the key is to connect them to real customer needs" (p.62)

Capabilities are those things that the company can do well repetitively. Production, logistics, daily human resource management, and partnerships—executing these day in and day out, handling the constant stream of issues that threaten to break these systems is an important capability for the company. Stalk (1992) points to the business processes that are established to insure that the system continues to work. He calls for strategic investments in the support infrastructure for these capabilities. Investments can only be strategic if the strategy aligns capabilities with competencies and resources as argued above.

The goal is to outperform the competition in the speed of response to customer needs, the consistency of the product specifications, an understanding of where the market is going and what it wants from its suppliers, and maintaining an agility to adapt to market and world changes (Stalk, 1992).

Given a specific set of resources, a company's capabilities allow it to apply those in an efficient manner. These enable continuous and uninterrupted operations. Improvements to existing processes, practices, and partnerships are part of these capabilities because they address incremental improvements to existing practices based on knowledge about those practices. They are an integrated part of operations, rather than being purposefully separated from operations. The competency to improve refers to the ability to see a product or process differently and to design the next generation that will replace it, not simply modify it.

"Core capability is an interrelated, interdependent knowledge system" (Leonard-Barton, 1992). These relationships limit progressive improvements to a rate and opportunity that can be accommodated within the entire current system, which differentiates them from competencies.

RESOURCES: THE SLEDGE

Sharif (1995 and 1999) suggests that there are four types of technology resources that are applied by a company that is innovating in its products and services (Figure 14.3). He describes these as:

- Technoware—object-embodied physical facilities
- Humanware—person-embodied human talents

- Infoware—record-embodied codified knowledge
- Orgaware—organization-embodied operational schemes

He also accepts that there are financial and natural resources avail-able which are not necessarily related to technology.

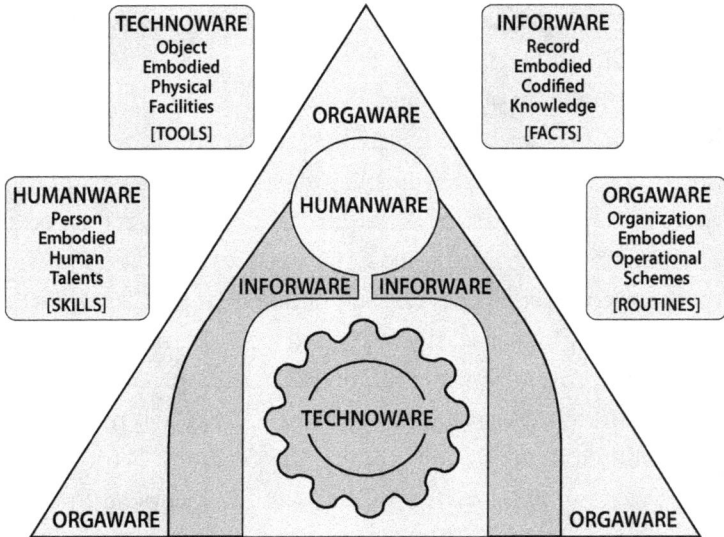

Figure 14.3. The technological resources available to a company fall into four major categories. Source: Sharif, 1995

Leonard-Barton (1992) suggests that there are 4 dimensions (or resources) that make up the knowledge set that enables capabilities and competencies. These are:

- Skills and Knowledge Base—knowledge and skill embedded in employees (i.e. Humanware)
- Technical systems—knowledge embedded in technical systems (i.e. Technoware)

- Managerial systems—formal and informal ways of creating knowledge (i.e. Orgaware and Infoware)
- Values and Norms—traditions from the founders (i.e. Orgaware)

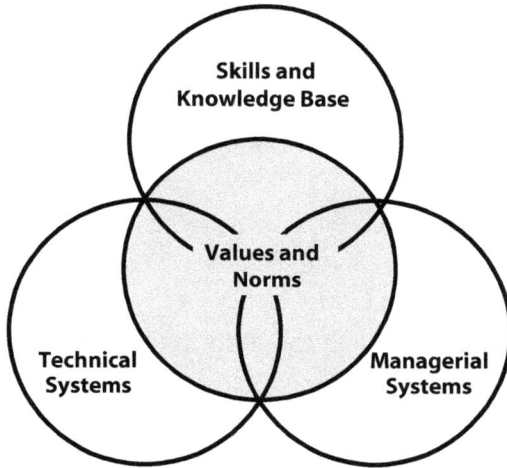

Figure 14.4. Leonard-Barton offers four dimensions of knowledge that contribute to organizational capabilities. Source: Leonard-Barton, 1992

The resources categorized by both authors are those that make-up the sledge behind the blade of the axe. These resources put weight behind organizational capabilities and the technological competencies that are penetrating and opening a market. Resources enable the organization to function.

INNOVATION ALIGNMENT STRATEGY

"From a competitive strategy point of view, technology can be used defensively to sustain achieved advantage in product differentiation or cost, or offensively as an instrument to create new advantage in established lines of business or to develop new products and markets." (Burgelman, 2004, p.143)

Burgelman also suggests that there are four dimensions of technology strategy:

- deployment of technology in the firm's product-market—position for differentiation (i.e. competencies),
- technology broadly applied across all activities of the firm's value chain (i.e. capabilities),
- resource commitment to various areas of technology (R&D) (i.e. resources), and
- use of organization design and management techniques to manage the technology function (i.e. daily operations).

These are consistent with the need to align competencies, capabilities, and resources in order to achieve significant and sustainable market penetration. In the introduction we also illustrated the effects of attacking the market with each one of these alone.

Attacking with resources alone refers to applying personnel, money, IP, or other assets without creating an organization that can deliver consistent products and services to companies. It also lacks the penetrating power of competencies to create new and innovative products that meet customer needs better than current offerings.

Attacking with competencies alone creates a prototype product and attempts to enter the market without the production, logistic, and marketing efforts required to consistently deliver the product or to make customers aware of its existence.

Attacking with capabilities alone creates a production and delivery system for a product that is mediocre and lacking innovative solutions to customer problems. This may succeed in placing yet another product on the shelves, but it will not significantly impact the market. This approach is most effectively used when the differentiating feature of the product is merely its price.

Aligning all three of these creates an organization and a product that can make a unique place for itself in the market, maintain its momentum, and grow its marketshare over time. Apple Computers is a strong example of this type of alignment. Their competency is in creating unique products that are differentiated in style and power from PCs. That competency continuously adds new innovations to existing products (i.e. iMac®) and creates entirely new products (i.e. iPod® and iPhone®). They back this up with the capability to produce the products with high quality and in sufficient quantities, accompanied by a marketing thrust that makes it clear why their offerings are unique and valuable. Behind these competencies and capabilities are the resources in personnel (humanware), technology (technoware), and organizational structure (orgaware) to source new products. They also possess unique values and norms that give everyone in the organization permission to think, act, and create differently. Google is another company that appears to have aligned its competencies in Internet search and data analysis, with its capabilities to deliver targeted advertising based on search results, and supports these with abundant human and financial resources. The company continues to create new products like Google Mail, Maps, Earth, Desktop, Toolbar, Blogger, Picassa, YouTube, Finance, Books, Shopping, Docs, Calendar, News, and 411 all of which build upon and extend their core competencies in collecting, analyzing, understanding, and selling data.

CONCLUSION

Prahalad and Hamel suggested that a company has only a few core competencies and emphasized the fact that corporate strategy must be built around these competencies. In this paper we build on those authors' ideas to create an innovation framework of competencies, capabilities, and resources which must work together to effectively penetrate the marketplace. This alignment is an essential part of the company's technology innovation strategy. If these three pieces are not aligned, then a competencies-based strategy will be ineffective because it will not be backed by the organizational processes or capabilities that are necessary to repeatedly carry those competencies to customers. Also, without sufficient resources, competencies and capabilities will be starved and unable to meet the demands of a market that has been penetrated. Initial successes will not persist long enough to capture a leadership position or to introduce subsequent waves of improved products and services.

ACKNOWLEDGEMENT

The author would like to thank Professor Nawaz Sharif at the University of Maryland for his assistance in clarifying some of the ideas in this paper.

References

Christensen, C. (1997). *The Innovator's dilemma: When new technologies cause great firms to fail*. Boston, MA: Harvard Business School Press.

Burgelman, Robert A., Christensen, Clayton M. and Wheelwright, Steven C. (2004). *Strategic Management of Technology and Innovation*, (4th Ed.), Chicago, IL: Irwin Publishers.

Hamel, G. and Prahalad, C.K. (1994). *Competing for the future*. Boston, MA: Harvard Business School Press.

Leonard-Barton, D. (Summer 1992). "Core capabilities and core rigidities: A Paradox in new product development". *Strategic Management Journal*, 13, pp. 111-126.

Porter, M. (1985). *Competitive Advantage: Creating and sustaining superior performance*. New York: The Free Press.

Prahalad, C. and Hamel, G. (May-June, 1990). "The Core competence of the corporation". *Harvard Business Review*, pp.79-91.

Prahalad, C. (May/June 1998). "Managing discontinuities: The Emerging challenges". *Research Technology Management*, 41(3), pp. 14-22.

Sharif, N. (1995). "The Evolution of technology management studies: Technoeconomics to technometrics". *Technology management: Strategies and applications for practitioners*, 2(3), pp. 113-148.

Sharif, N. (1999). "Strategic role of technological self-reliance in development management". *Technological forecasting and social change*, 44(1), pp. 219-238.

Stalk, G., Evans, P., and Shulman, L. (March-April 1992). "Competing on capabilities: The New rules of corporate strategy." *Harvard Business Review*, pp. 57-69.

The list of Google applications appearing in this chapter are all registered trademarks of Google Inc.

Originally Published in *Research Technology Management*, Sept-Oct 2008

CHAPTER 30

THE INNOVATION-CENTRIC COMPANY

INNOVATION MANAGEMENT

The ability to manage innovation successfully is one key to the success of companies like Dell, GE, and Microsoft. In some cases, innovation is synonymous with research. In others, it is the application of existing ideas to new problems. GE's research labs practice innovation when they create new synthetic materials that can become part of medical implants. Microsoft practices innovation when it recognizes the importance of search technology and engineers it into the operating system. In both cases, the companies recognize that the future is not a copy of the past and they must take action to insure that they are a prominent player in the new structure of things.

Every company faces a unique set of challenges when gearing itself up for innovation. These challenges stem from the current position in the market, historical processes, internal capabilities, leadership support, and available budgets.

SPECIFIC MANAGEMENT CHALLENGES

To whom should the responsibility for innovation fall? It cannot be the responsibility of everyone in the organization because that would leave no one to handle all of the other operations of the company. But it cannot be treated as an island separated from the rest of the company either. Innovation will involve a new way of thinking for people across the organization. It is something that must be supported and communicated from the top of the organization to the bottom. Is an executive level necessary? Can corporate innovation be accomplished without the direct support of a corporate executive?

Who does innovation? If everyone will be the recipient of new processes, equipment, and tools, then who is responsible for creating these and identifying the best application? Is a research laboratory the best environment for creating innovation that will be applied within the company? Or are the internal halls of the company the best source of improved ideas?

Must a company choose between innovation and stability? Can a company provide employment, process, and cultural stability to meet the security needs of employees, while at the same time innovating to remain at the head of the industry? Constant change is destabilizing and demoralizing to a large number of people. They seek jobs that are secure, stable, and repetitive. Many people need to be able to master the processes they are responsible for, rather than operating in a changing environment that they will never master or even fully understand.

Must a company be innovative alone? Should innovation be carried out internal to the company and shielded from the eyes of everyone else in the industry? Or can a company form unique partnerships that will share in innovative changes? In the latter case, the sustaining advantage of innovation comes from the inability of competitors to duplicate the combination of skills and the processes themselves.

INNOVATION OFFICE

To become masters of innovation, organizations must make some basic changes to their structure of responsibility. The first of these is to establish an Innovation Office. This organization is charged, not with originating all innovations, but with motivating, tracking, managing, and measuring innovation across the company. Intel Corporation has recently created the corporate position of Chief Technology Officer to assist in this process. Patrick Gelsinger, the first corporate-wide CTO, is responsible for overseeing the work of the research and laboratory organizations. He is expected to keep those organizations focused on creating products that are aligned with the company's strategic vision and mission—to become a hub in which computing and communications technologies can merge to create products that are radically more powerful than those that exist today. Gelsinger's organization is not responsible for marketing Pentium chips or pressing into the cellular telephone market. Instead, the CTO office must strive toward a future in which an Intel product is at the center of new devices and capabilities that include both communication and computation. The integrated cell phone and PDA is an early version of these devices, but certainly not the culmination of the vision.

INNOVATION AS CORE

Companies have historically considered the product to be the core of their business. This has been followed by "service as core" and "process as core". A product-centric company may be something like a mid-20th century General Motors Company. The core of their business was focused on the production of the automobile. Everything within the company was structured to insure that the automobile contained the performance, style, quality, and price that met the customer's needs.

A service-centric company has a broader perspective in which it may provide a product, but it also sees itself as meeting larger needs of its customers. A service-centric automobile company would not limit its offerings to the automobile, but would expand these to include more of the customer's needs. An automobile-buying customer requires financing to be able to afford the product and that financing includes a profit margin. Therefore, the late century GM also provided financial services, insurance, title application, license transfer, and any other services that make it easier for the customer to buy an automobile while also providing additional profits for the company.

A process-centric company may see the customer as a person who has a constant, life-long need for transportation. That person needs to buy an automobile along with the service components. But they also need transportation when the automobile is being repaired. Since repairs can be done at any number of shops, the GM dealer must provide an incentive to bring the customer to the dealership for repairs. One of those incentives is the loaner car and the shuttle service. These insure that the owner of a GM automobile has a transportation solution at all times during the life of the vehicle. It also builds a relationship that is designed to bring the customer right back to the dealer for the next automobile purchase, preferably a lifetime of purchases.

In an innovation-centric company, the goal is to meet the needs of the customer today and those that have yet to be imagined in the future. A company must demonstrate that their innovation moves them from a customer's current needs to their future needs before the customer gets there. Customers will learn which companies can only satisfy today's problems and which are already imagining and solving the problems they will have in the future. The innovation-centric company is establishing itself as a lifelong partner. GM's OnStar® system can be cast in this innovation-centric light. As people find themselves more independent and disconnected from each other, they learn that they can no longer count on other motorists to render assistance. Therefore, OnStar® is a GM innovation that meets the needs of a large part of the automobile customer-base before the customer's realize that they need it. Like cell phones, the item moves from luxury to necessity as customers catch up with the innovation that has already been done by GM.

Companies that are serious about innovation must focus themselves around the needs of the future. They must tie the organization to the source of innovation and structure the company such that they are extracting the maximum value from innovation as quickly as possible.

FLEXIBLE WORKFORCE

Thriving and surviving in an innovation-centric company is not something that is natural and easy for many of today's employees. Creating a culture and a workforce that is functional and motivated in such an environment is a major undertaking. Both the company and the employees must build a relationship that grows stronger because they know how to evolve and change together. The company must be able to teach employees to thrive in the new environment and the employees must be willing to trade old behaviors for new. Because this transition is difficult, employees who make the change should be more valuable to the company. Perhaps a relationship based on flexibility, adaptation, and innovation can become the foundation for lifelong employment. McFarland Strategy Partners teaches companies that, "The best way to grow a business is to grow the people. It just doesn't make sense to be replacing people all the time." (Kurtz, 2004)

In the 19th and 20th centuries labor struggled to build a foundation for stable employment based on unchanging responsibilities and processes. Someone who could master a specific step in the production process was more valuable as long as that step never changed. In the 21st century that stability may be based on flexibility. The value to the company is not in the person's ability to become a master at one task. But rather they are valuable because they have the ability to flex, grow, and change to master a wide variety of tasks placed before them. Though they may never master one job, they are able to become competent at many successive jobs. People with this flexibility in an innovation-centric company cannot be eliminated because flexibility coupled with competence is too valuable to waste.

INNOVATION PARTNERSHIPS

As described earlier, innovation and the mastery of technology does not respect organizational or international boundaries. An innovation-centric company cannot limit itself to people and processes that reside within a specific region, facility, or company. Leaders in innovation must be able to draw innovation into synergistic relationships from any number of sources around the planet.

Managing innovation will be synonymous with managing multi-company teams that cross geographic and international boundaries. They are not bound together by their physical addresses, community history, or even professional specialty. Instead they are bound together by their faith in a vision of the future and their eagerness to create something that does not yet exist. In such an environment, the issues like "not invented here" are less an issue because everyone is looking at how the invention can be applied or how it leads to something new that will be invented here. When the allegiance is to a vision or ideal, then issues of company origins and physical location are not important as long as the partnership can enable progress better than a single in-house effort.

Innovation-centric companies require innovation-centric people. Innovation management is not about controlling and standardizing people and processes, but about enabling and optimizing them. The responsibility of the manager is to cultivate, train, recruit, and combine people to continually improve the innovation process of the combined partnership.

References

Chesbrough, H. (2003). *Open innovation: The New imperative for creating and profiting from technology*. Boston, MA: Harvard Business School Press.

Christensen, C. and Raynor, M. (2003). *The Innovators solution: Creating and sustaining successful growth*. Boston, MA: Harvard Business School Press.

Hargadon, A. (2003). *How breakthroughs happen: The Surprising trust about how companies innovate*. Boston, MA: Harvard Business School Press.

Kurtz, R. (June 2004). "Testing Testing …". *Inc. Magazine*.

Originally Presented at *University of Maryland*, October 2006

CHAPTER 31

DISRUPTION, GLOBALIZATION, AND INNOVATION MANAGEMENT

TECHNOLOGY IMPACT

Management literature is filled with examples of the changes wrought on business by the addition of technology and the creation of entirely new businesses whose primary product or service is technology. These changes have become of vital interest to companies who see technology as a means to improving their operations. They are also important to the inventors and vendors of this technology, encouraging them to identify the next product or service that customers will flock to. The term "killer app" is often used to describe the most wildly successful technologies or products. In the past, such killer apps have been the personal computer, laptop computer, personal digital assistant, windowed operating system, word processor, Internet, electronic mail, World Wide Web, web browser, information portal, cellular telephone, and wireless messaging. Each of these has been so widely adopted that the leading providers have risen to the pinnacle of commercial success. Companies like Microsoft, Dell Computers, Palm, Research in Motion, Yahoo, Google, and Nokia have all ridden from obscurity to international prominence on the back of a killer app.

THREE MAJOR TRENDS

The power that technology brings to an organization has the potential to launch new ventures to global prominence and overturn established incumbents. Therefore, success in implementing or creating new technologies has become a topic of intense interest. Below are three major trends in the integration of technology and information into an organization.

1. Disruptive Impact of Technology

As technology has become a new tool for executing business functions, the optimization of these capabilities has opened a new domain for companies to achieve a competitive advantage in their industry. The ability to identify current sales and inventory levels has led to real-time supply systems. Information on credit card transactions contains clues that can identify fraudulent usage in near real-time. The military's ability to access real-time information about enemy movements and positions from anywhere in the world allows them to control all major confrontations.

Christensen (1997) illustrates the early emergence of technology transformation in earth excavation equipment from the 1920's to the mid-1950's. In the 1920's, only cable driven shovels possessed the power and leverage necessary to meet the needs of major construction projects. In 1947 the first hydraulically powered shovel was introduced in Britain and rapidly copied in America. This equipment was called the "backhoe" because it was small enough to attach to the back of farm equipment. The hydraulic seals in the equipment could support only a ¼ cubic yard bucket. Therefore, the backhoe was relegated to small farm and residential projects and posed no threat to the large construction jobs handled by cable driven shovels that could extract 4 cubic yards of earth in a bucket.

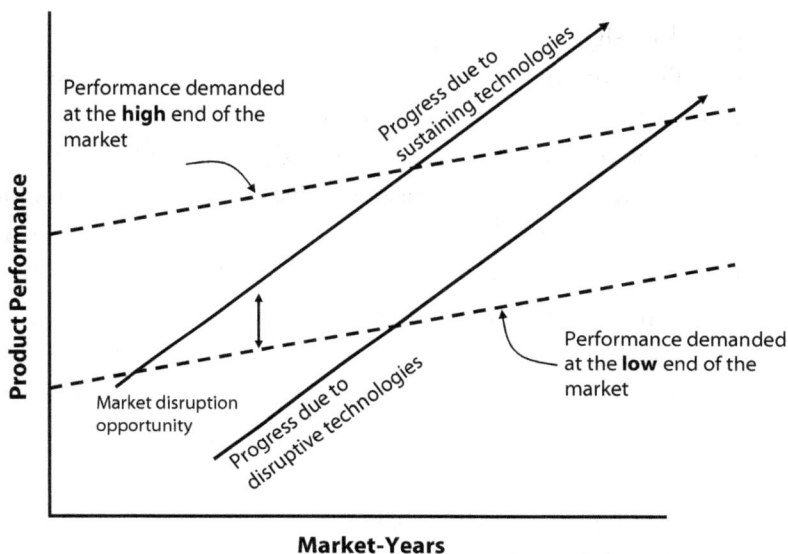

Figure 16.1. Disruptive Technological Change (Christensen, 1997)

However, advances in seals and pumps soon led to hydraulic equipment that could handle larger buckets. By 1965, hydraulics could excavate 2 cubic yards per bucket and by 1974 over 10 cubic yards. These advances made hydraulics more powerful than the traditional cable powered excavators. This "disruptive innovation", as Christensen terms it, totally destroyed the cable excavation equipment industry and replaced it with a new group of companies that had developed hydraulic equipment. Christensen maintains that such technological advances will generate similar disruptions in every industry and he illustrates this with in-depth analysis of the steel and disk drive industries. The crux of Christensen's argument is illustrated in Figure 16.1. Upstart competitors can enter a market by serving low-profit margin customers. Because entrenched companies are more motivated to serve high-profit customers, they often cede the low-end to the upstart company. However, in an industry where technology can be applied to significantly improve product performance, the new entrant will move up market and attack the entire customer base of the entrenched company. This has occurred in a number of industries and has drawn the attention of business leaders around the world.

In *The Innovator's Solution*, Christensen lists a number of technologies and companies that have overturned established industry leaders. These include Amazon.com, Bloomberg LP, Canon, Cisco, eBay, Federal Express, Intuit, Google, Intel, and RIM (Christensen, 2003). Each of these used emerging technologies to gain a small position in the market and then relied on the advancement of their technology to elevate them past long-entrenched leaders, significantly disrupting the industry.

Industry competition is no longer dominated solely by historical presence or financial size. The rapid advancement of technology is a lever that can be used to enter a market and move to a dominant position.

2. Technology Globalization

The second trend is the emergence of global competence to provide competitive products and services. Though a technology may be born in the United States, it can easily migrate to any country that is prepared to adopt it and invest in improvements and applications. The transistor was invented at Bell Labs and RCA spent millions attempting to improve its properties to replace vacuum tube radios. Sony, on the other hands, turned early transistors into small pocket radios with inferior sound quality. That product gave them a position in the radio market, but a low profit margin position against which RCA did not want to compete. However, transistors proved to be the future of all radio equipment and Sony was able to ride its position in transistor radios to market dominance and to completely drive RCA out of the industry (Chandler, 2001).

Later in the century, the explosion of information technology in business led to the creation of a new type of service—the IT consultancy. Companies like Electronic Data Systems, Accenture, Bearing Point, and PricecoopersWaterhouse developed large IT service businesses based on those needs. But, just as technology-based products migrated from America to Japan in the 20th century, technology-based services are migrating to more affordable countries in the 21st century. These types of jobs require special knowledge and expertise, but in a form that is repetitive and teachable. The ability of entrepreneurs and universities in India, China, Puerto Rico, and Russia to train such people and organize businesses to provide these services at lower costs has led to significant outsourcing of American IT jobs. Business Week estimates that there are currently 150,000 IT engineers in Bangalore, India and that one third of all new IT development work by U.S.-based companies is being done in India (Kripalani, 2003). This transition is almost identical to that of the electronics industry from 1950 and 1980. The standardization of computer equipment and software interfaces, improved global telecommunications, and competitive educational programs provide all of the ingredients necessary to perform software development, IT back office services, and call center support from anywhere on the globe. Just as the Japanese were able to take over the electronics markets, the Indians and Chinese are in a position to take a major share of the IT services market. Consulting firm A.T. Kearney estimates that by 2008 over 500,000 financial service jobs will be outsourced to companies outside of the U.S. (Kripalani, 2003).

More recently the trade press is emphasizing that some of these jobs are being pulled back into the United States. But, this is a temporary balancing measure driven by the fact that companies outsourced too much capability too soon. The IT services industry in emerging countries is using Christensen's disruptive innovation curve to attack the market at the low end. Over time their expertise will increase and the entire IT service industry could be lost to India and China just as the entire electronics industry was lost to Japan in the last century.

Technology and information are tremendous leveling tools. They do not discriminate based on political ideology, ethnic origins, language, social class, geographic location, or corporate boundaries. Companies and countries that develop an internal, integrated learning base possess a competitive advantage that will allow them to dominate an industry (Chandler, 2001).

3. Innovation Management

The third trend that we present is the growing importance of innovation management. Both the companies that create and those that apply technology need a new form of management—one that optimizes an organization in motion rather than an organization at rest. Managers who have mastered the skills necessary to keep a production line operating or a telephone switch working, have not necessarily mastered the skills necessary to successfully evolve those operations. Ongoing business operations are usually about repetitive activities. They call for methods to systematize activities that are performed over and over every day. The process is static and the personnel are molded to fit the process.

Conversely, the management of innovation requires a process that is constantly changing and keeping people flexible so they can perform well in different situations every year. A static process cannot be defined to meet these needs and people cannot be molded into a single form. Instead, the manager must find methods to measure and improve performance, but allow the people to adapt their performance to emerging needs. Innovation Management is a way of empowering people, setting guidelines, and building networks that can drive a business to innovate and reinvent itself in the midst of a changing market.

Impact on Management Practice

These changes have had significant impacts on management practice. First, most functional organizations have had to find a way to integrate technology professionals into their operations and their professional communities. The earth excavation field can no longer be populated only by those who operate excavators. This community must now include people who create tools to make excavation more efficient. It must find a way to welcome people with entirely different skill sets and unique perspectives on a problem.

Valuable resources are no longer limited to the four walls of the company's facilities. A team may include people throughout the company, within partner organizations, and across national boundaries. A team is less likely to be a homogeneous group and more likely to require an appreciation for the perspectives, mores, and traditions of a much broader set of people.

The executive ranks had previously been dominated by three major titles, the CEO, COO, and CFO. Together they determined the direction of the company and assumed responsibility for its financial health, operational capability, and service to shareholders. But as technology has become a major ingredient in all operations, the executive ranks have been joined by others with special responsibilities for internal IT, product research, corporate knowledge, and information security. Companies now count the CIO (information), CTO (technology), CKO (knowledge), and CSO (security or strategy) as important members of the executive team.

INNOVATION MANAGEMENT

Technical applications that are propelling business, government, and military operations are developed by creative, intelligent, and visionary individuals and organizations. To some degree, the velocity of a significant part of the economy rests on the shoulders of these visionaries.

The speed with which they innovate has a direct impact on the ability of their customers to integrate these new tools and improve their business performance. Opportunities that are lost in these technology companies represent potential lost improvements across a large sector of the world economy. Therefore, the optimization of innovation and creation is an important economic and business topic. The ability to innovate and create is essential, but the fruits of that ability cannot be realized without the ability to manage that innovation. To quote Henry Chesbrough, "technology by itself has no single objective value. The economic value of a technology remains latent until it is commercialized in some way" (Chesbrough, 2003).

The ability to successfully manage innovation is one key to the success of companies like Dell, GE, and Microsoft. In some cases, innovation is synonymous with research. In others, it is the application of existing ideas to new problems. GE's research labs practice innovation when they create new synthetic materials to improve medical implants. Microsoft practices innovation when it embeds search technology into its operating systems. In both cases, the companies recognize that the future is not a copy of the past and that they must take aggressive action to remain in their dominant position.

High Stakes

Alfred North Whitehead said, "The greatest invention of the nineteenth century was the invention of the method for invention" (Buderi, 2000). The 19th century saw a major transformation away from a reliance on independent scientists and inventors as the source of innovation and toward the creation of internal research labs and product development teams. Buderi (2000) describes the emergence of the German organic dye industry from the explorations of individual scientists to the establishment of corporate research staffs under the oversight of brilliant scientists. Companies like Perkins, Bayer, BASF,

Höchst, and AGFA all emerged in the 19th century and built their entire business on the products of their research laboratories.

In the 20th century, RCA enjoyed a near monopoly on radio technology and rode that position from 1919 to 1980. But business historian Alfred Chandler attributes RCA's downfall in the industry to its lack of an "internal learning base". The company controlled the critical patents for radio technology, but did not have the in-house expertise to develop new products and push that technology into new markets. Therefore, it was vulnerable to Japanese companies like Matsushita and Sony as they created new electronic products based on the transistor (Chandler, 2001).

Today, the same effect can be seen in computers, software, and Internet services. The world's preferred search engine shifts as new competitors release better technologies and word of it spreads like wildfire through the Internet. Search engine leaders like Alta Vista and Yahoo were completely eclipsed when Google emerged from a Stanford research project.

Objectives

The primary objective of innovation management is to arrive at marketable products or services before a competitor does. As Chesbrough pointed out, it is the commercial application of technology, not just the creation of technology that defines successful innovation. James McGroddy, the head of the IBM T.J. Watson Research Center, realized the importance of this connection as early as 1989 when he observed that the research center had five Nobel Prize winners on its staff, but "research was not making much difference to the company" (Buderi, 2000).

New technology is born with great potential. In fact, the promises inherent in a technology always exceed society's ability to absorb, adopt, or apply it (Chakravorti, 2003). Innovation management seeks to identify that portion of the technology that is most amenable to the current social and market situation. It seeks to push forward applications that might become the next killer app and maintain the company's leading edge or propel it to the front of the pack at the expense of the current leader.

Action

Most academics and practitioners are familiar with the story of the Xerox Palo Alto Research Center (PARC), which, though very successful at creating innovations for the photocopier industry, lost its lead and its own technologies in the emerging computer industry (Smith, 1999 and Buderi, 2002). In the wake of this missed opportunity, companies are seeking new business models to insure that they capture the full value of their own innovations.

Starting Innovation. The first step in the process must always be starting the process of innovation. Any organization with a sufficient budget can create an innovation or research department. But that is not sufficient for succeeding at the process. Markides (2002) maintains that beginning this process requires redefining the business. If the company's definition of itself remains the same, there is no place for innovation, change, and unique action. He recommends that a company redefine the answers to the following questions:

- Who is our customer?
- What products or services are we offering these customers?
- How can we offer better products or a new way of doing business with our customers?
- What are our unique capabilities?

After beginning with these questions, a company is ready to move on to establishing a management process for driving innovation.

The Best People. In considering an invitation to join the new Microsoft Research in 1993, Linda Stone's reaction was, "Wow. I could come in to work every day and work with this group of people" (Buderi, 2000). It was the amazing group of leading researchers that compelled her to accept the offer and put aside her previous opinions of the Redmond, Washington company. Great innovation requires great people. The staff of the government's Manhattan Project, Lockheed's Skunk Works®, and Apple's Macintosh® computer all drew upon the very special abilities of carefully selected people. Warren Bennis studied these and other projects and created a list of requirements for organizing genius to achieve great objectives. His list of 15 characteristics is dominated by the need for great people. In his words, the need is for "superb people", a "strong leader", "mission focus", talent that can work together, and "optimism" (Bennis, 1997).

The right processes, equipment, and mission cannot compensate for the wrong set of people. In some cases, the need is for technical knowledge, in others creativity, the ability to organize, and the gift of motivation.

Tenets of Innovation. Arun Netravali, former president of Bell Labs, maintains that all successful research follows three tenets—speed, complexity, and cannibalization (Buderi, 2000). Research must move fast in its pursuit of advances. Reaching the goal one month or one year after the competition is almost as bad as not reaching the goal at all. That work may lay important groundwork for the next innovation, but it will have lost most of its commercial value to its faster rival.

In the real world there are no ideal laboratory conditions. The unique value in an innovation often stems from its ability to handle the complexity that exists in business operations. Computer switching equipment from Lucent must be able to deal with a wide variety of signal speeds and computer systems from a number of vendors. An innovation for the fastest switch for a single signal and computer combination is not nearly as valuable as a less than optimal switch that can work effectively with all combinations of signals and equipment in the network.

Finally, a research organization must obsolete its own products, services, processes, and organizational structures. It must push for innovation and improvement even at the expense of its existing products and services. If an organization does not cannibalize its own products, then a competitor will. Intel Corporation is famous for this type of behavior. The last version of their CPU is just becoming a market staple when the company releases the next generation that significantly surpasses it. But, if Intel did not do this to its own products, AMD would. Cannibalization and speed are necessary partners in the innovation process.

Value Extraction. At Xerox PARC in the 1980's the value of all research was measured by the degree to which it contributed to a complete system that could be sold into a client organization. Up to that time, this process had served it well in creating new technologies for its photocopier products. However, as its research created technologies more useful in computing and communications, it was the wrong metric for identifying the value inherent in the technology itself. As a result, a number of researchers left Xerox to start their own companies and were able to license the technology for a small price. Xerox hoped that the success of those technologies would stimulate sales of their existing products and, therefore, did not seek major ownership positions in the spin-offs.

This process led to the creation of over 20 new companies focused on computer hardware and software. The most successful of these have been 3Com and Adobe, followed by a number that were reacquired once they had proven their value in the market (Chesbrough, 2003).

Viewing this as a lesson in value extraction, major players like IBM, Intel, and Lucent have created new methods for managing technologies from their laboratories. Like Xerox, IBM initially viewed itself as a complete system provider and sought to develop technologies that would improve its major systems. But, in recent years it has changed that model and begun to extract value in different ways. Its first major change was to become a system integrator of standard computer components—the birth of the IBM personal computer. From there it moved into IT services that were not limited to IBM computer products. These services competed directly with EDS and consulting firms like Accenture and Bearing Point. More recently it has decided to extract value through technology licensing. Its 2½" hard drive technology has become a staple in a number of vendors' laptop computers. Though the technology was originally developed to give a competitive advantage to IBM's ThinkPad® line of laptops, the company has sold as many hard drives to other vendors as it has sold ThinkPad® laptops. This has doubled the company's revenues for the drives. IBM has begun licensing its intellectual property and allows other companies to build hardware based on its unique IP. Finally, IBM has started a program entitled First of a Kind (FOAK) in which it explores a new and unproven technology in conjunction with a major customer. Both participants contribute their own resources and successful technologies become IBM IP, which supports future innovation (Chesbrough, 2003). This value extraction is considerably more complex than the typical 1980's era practices at PARC, Watson, and Bell labs.

Intel spent $3.8 billion on innovation in 2001. With such a large investment, the company must do everything it can to extract the maximum value from this innovation. One extremely successful practice has been their "Copy Exactly" policy. Chip development and production facilities are constructed as exact copies of each another. Both use the same machinery, models, and configuration settings to produce their chips. The layout of the production line is identical as are the processes for operating the system. This insures that the newly developed chips and the production line chips will have exactly the same characteristics. The productivity levels in the development facility are identical to the productivity levels that will be achieved in production. This process has saved significant time and cost in selling customers, receiving production approvals, and migrating new processes and equipment into the production environment.

Intel also improves their value extraction by establishing research partnerships with universities and consortia and then managing those closely. Contrary to the more common practice of giving money to a university and allowing it to pursue its own path, Intel assigns engineers to work directly with the students so they have the advantage of the most current level of knowledge and can advance more quickly. Intel has also established a venture capital organization that funds external companies—which it works closely with to improve their productivity and effectiveness.

Organizational Process. The actual organizational process for an innovative group appears to be less unique than the characteristics described above. These processes are defined in a number of texts, one of these being *Radical Innovation* (Leiffer et al, 2000). With the cooperation of ten major research organizations, including GE, GM, IBM, and DuPont, the authors attempted to identify a process that has been most effective for managing innovation and turning innovative ideas into products. The result was a list of the following essential activities.

1. **Market Learning.** Instill a mindset of constant market learning. Use large, small, formal, and informal sampling to extract information. Be open to counter-intuitive information. Tap into User Innovation Communities. Examples of these include the Open Source software movement, homebrew computer makers, kit

airplanes, telephone "phreaking", computer virus underground, amateur game developers, and "monster garage" mechanics (von Hippel, 2002 and Thomke, 2003)

2. **Find a Place in the Value Chain.** Map technical innovation to business success. Every innovation must demonstrate that is has value to the company. The value extraction practices of the IBM T.J. Watson Research Center described above illustrate this process very well.

3. **Effectively Build Business Models.** Understand the value chain and how the company plays in that chain. Modify the business model to incorporate innovative products. Prepare for war with established business units.

4. **Fight for Funding.** Every innovation team should include an expert in resource acquisition. Senior management must balance investments in innovation against investments in current operations. Innovation projects must win these decision-makers to their side. Seek sources of external funding.

5. **Build Partnerships.** Modern products and services often require more resources than exist within a single organization. Build partnerships with other companies without jeopardizing the internal value of the innovation.

6. **Transition to a Business Unit.** Establish a transition team and write a transition plan. Recruit senior managers as transition champions. Invest in transition, just as you did in innovation. Successful transition is a key part of successful innovation.

7. **Managing Individuals.** Balance the composition of the innovation team. Include people with skills beyond intelligence. Retaining talented people in a risky environment requires investment and long-range planning. Contract for specialized help outside of the team. Transition personnel as you transition innovation.

This process has proven useful at a number of large organizations. Small companies may find it difficult to implement or finance. But, the flowering of the venture capital community has made it possible for even small start-ups to deploy significant capital toward their innovation process.

CONCLUSION

Technology and information management is much more than just installing, optimizing, and operating IT systems within an organization. Technology has become a central feature of nearly every business and has the power to significantly influence its market success. Effective management of the innovation process is an essential step in structuring and building a company in the 21st century. Without success in this area, competitors will move aggressively through the market and displace companies that are not innovating, that are standing still as Alta Vista and RCA stood still in the 20th century.

References

Bennis, W. 1997. *Organizing genius: The Secrets of creative collaboration.* Reading, MA: Addison-Wesley.

Buderi, R. 2000. *Engines of tomorrow: How the world's best companies are using their research labs to win the future.* New York: Simon & Schuster.

Buderi, R. June 2002. Mhyrvold's exponential economy. *Technology Review* online. http://www.technologyreview.com/articles/print_version/qa0602.asp

Chakravorti, B. 2003. *The Slow pace of fast change: Bringing innovations to market in a connected world.* Boston, MA: Harvard Business School Press.

Chandler, A. 2001. *Inventing the electronic century: The Epic story of the consumer electronics and computer industries.* New York: Free Press.

Chesbrough, H. 2003. *Open innovation: The New imperative for creating and profiting from technology.* Boston, MA: Harvard Business School Press.

Christensen, C. 1997. *The Innovators dilemma: When new technologies cause great firms to fail.* Boston, MA: Harvard Business School Press.

Christensen, C. and Raynor, M. 2003. *The Innovators solution: Creating and sustaining successful growth.* Boston, MA: Harvard Business School Press.

Kripalani, M and Engardio, P. December 8, 2003. "The Rise of India". *Business Week* online.

Leifer, R, et.al. 2000. *Radical innovation: How mature companies can outsmart upstarts.* Boston, MA: Harvard Business School Press.

Markides, C. 2002. "Strategic Innovation". in *Innovation: Driving product, process, and market change.* Edited by E. Roberts. San Francisco, CA: Jossey-Bass.

Thomke, S. 2003. *Experimentation matters: Unlocking the potential of new technologies for innovation.* Boston, MA: Harvard Business School Press.

Smith, D and Alexander, R. 1999. *Fumbling the future: How Xerox invented then ignored the first personal computer.* New York: HarperCollins.

von Hippel, E. 2002. "Innovation by user communities: Learning for open-source software", in *Innovation: Driving product, process, and market change.* San Francisco, CA: Jossey-Bass.

Originally Submitted to *Technovation*, May 2006

www.ingramcontent.com/pod-product-compliance
Lightning Source LLC
Chambersburg PA
CBHW061207220326
41597CB00015BA/1542